Lab Pack

for

Essentials of PHYSICAL GEOGRAPHY

Seventh Edition

Lab Pack

for

Essentials of PHYSICAL GEOGRAPHY

Seventh Edition

Robert E. Gabler
Western Illinois University, Emeritus

James F. Petersen
Southwest Texas State University

Michael L. Trapasso
Western Kentucky University

Australia • Canada • Mexico • Singapore • Spain • United Kingdom • United States

Printed in the United States of America
1 2 3 4 5 6 7 07 06 05 04 03

Printer: Patterson Printing Company

ISBN: 0-03-033868-9

For more information about our products, contact us at:
Thomson Learning Academic Resource Center
1-800-423-0563

For permission to use material from this text, contact us by:
Phone: 1-800-730-2214
Fax: 1-800-731-2215
Web: http://www.thomsonrights.com

Cover image: Yosemite National Park by Kerrick James, Getty Images.

Brooks/Cole-Thomson Learning
10 Davis Drive
Belmont, CA 94002-3098
USA

Asia
Thomson Learning
5 Shenton Way #01-01
UIC Building
Singapore 068808

Australia/ New Zealand
Thomson Learning
102 Dodds Street
South Street
Southbank, Victoria 3006
Australia

Canada
Nelson
1120 Birchmount Road
Toronto, Ontario M1K 5G4
Canada

Europe/Middle East/South Africa
Thomson Learning
High Holborn House
50/51 Bedford Row
London WC1R 4LR
United Kingdom

Latin America
Thomson Learning
Seneca, 53
Colonia Polanco
11560 Mexico D.F.
Mexico

Spain/ Portugal
Paraninfo
Calle/Magallanes, 25
28015 Madrid, Spain

PREFACE

The exercises contained in Lab Pack were developed by the authors to both complement and supplement information presented in Essentials of Physical Geography, seventh edition. As the Table of Contents details, there are 28 major units within Lab Pack which contain a total of 58 different exercises. The exercises vary in length and difficulty but all were designed to help you achieve a greater understanding and appreciation of physical geography.

Please note that the exercises require that you provide certain materials – rulers, colored pencils, etc. These materials are clearly stated at the beginning of each exercise. Further, because these exercises are based upon textbook material, it is imperative that you read the assigned matter before undertaking the exercises and have your textbook with you while you work on them.

Acknowledgments

Several of the exercises in Lab Pack were modeled after exercises developed in the Department of Geography at Western Illinois University and the Department of Earth Science at Pierce College, Tacoma, Washington. The authors would especially like to thank Joanne R. Shelley (Pierce College) for materials and assistance in Units 18, 19, 21, 22, and 28; and Douglas Miller (Western Illinois University) who wrote the exercises contained in Units 4, 5, 16, and 17.

TABLE OF CONTENTS

Name ______________________________________ Instructor ____________________

Date ______________________________________ Section Number _______________

UNIT 2 - Locational Systems and Time Zones

For use with Chapter 2

2.1 Latitude and Longitude; Public Lands Survey

Reference Pages in Text: Chapter 2, pp. 32 - 40

Materials Needed:
Student Supplied: pencil
Instructor Supplied: atlas and globe

Purpose: This exercise will help you become familiar with the global grid system (latitude and longitude) and the Public Lands Survey System as a means of locating places.

A. Latitude and Longitude

1. What cities are located at the following grid coordinates?
   a) 40°N, 75°W
   b) 30°N, 90°W
   c) 34°S, 151°E
   d) 41°N, 112°W

2. What are the grid coordinates of the following cities?
   a) Portland, Oregon
   b) St. Petersburg, Russia
   c) Rio de Janeiro
   d) Your home town

3. The point on Earth's surface opposite 52°N latitude, 140°W longitude is

4. You are located at 10°S latitude, 10°E longitude; you travel 30° north and 30° east. What are your new geographical coordinates?

5. You are located at 40°N latitude, 90°W longitude. You travel due north 40°, then due east 60°. What are your new geographical coordinates?

B. Public Lands Survey System

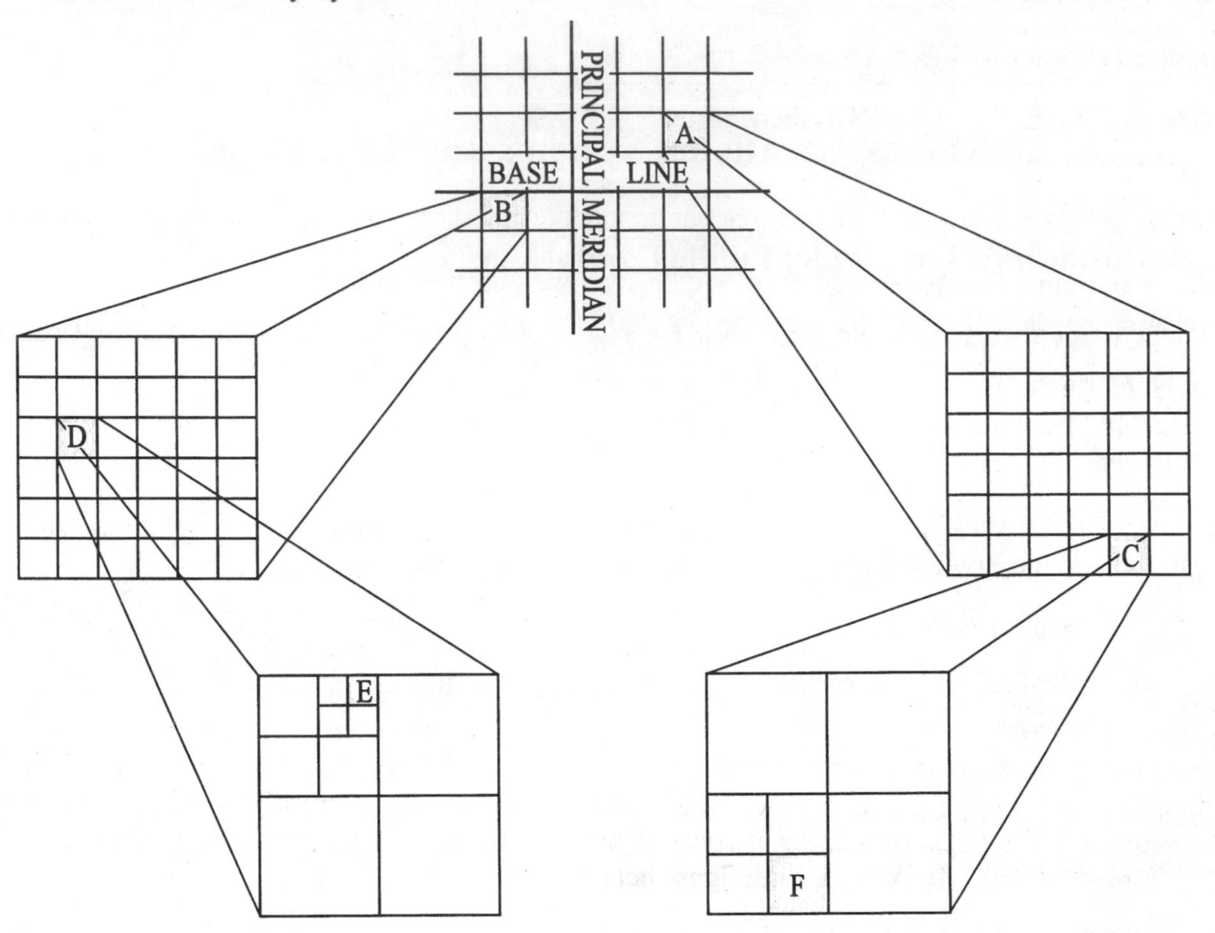

1. Based upon the diagram above, answer these questions:
   a) What is the Township and Range of:
      Location A?
      Location B?

   b) What is the Section Number of:
      Location C?
      Location D?

   c) What is the complete Public Lands Survey System description of:
      Location E
      Location F

   d) How many acres are contained within:
      Location E?
      Location F?

2.2 Time Zones

Reference Pages in Text: Chapter 2, pp. 37 - 38

Materials Needed: Student Supplied: pencil
Instructor Supplied: atlas

Purpose: This exercise will help you better appreciate and understand the time zones of the world.

A. Use Figure 2.8 (p. 37) and an atlas to answer the questions below.

You are located in New York City where it is 5:00 p.m. Tuesday, October 15. What time and day is it in:

1. Denver, CO
2. Hilo, HA
3. New Orleans, LA
4. Frankfort, Germany
5. Tokyo, Japan
6. Sydney, Australia

B. Time zones are based upon the fact that, if viewed from the North Pole, Earth rotates counterclockwise through 15 degrees of longitude each hour ($360 \times 24 = 15$). Consequently, places to the west of a specific location have earlier times and places to the east have later times than that location (see the figure below).

Early navigators also used this fact to determine their longitude. On board every ship was a chronometer–a very accurate clock set to Greenwich Time (remember Greenwich is on the prime meridian), and a sextant which is used to determine the exact moment of solar noon.

If it were solar noon at your location and the chronometer on your ship indicated it was 8:00 p.m. in Greenwich then you would have to be 120 degrees from the prime meridian ($8 \times 15 = 120$). Since your time is earlier you must be to the west of Greenwich–so you are located at 120° W longitude.

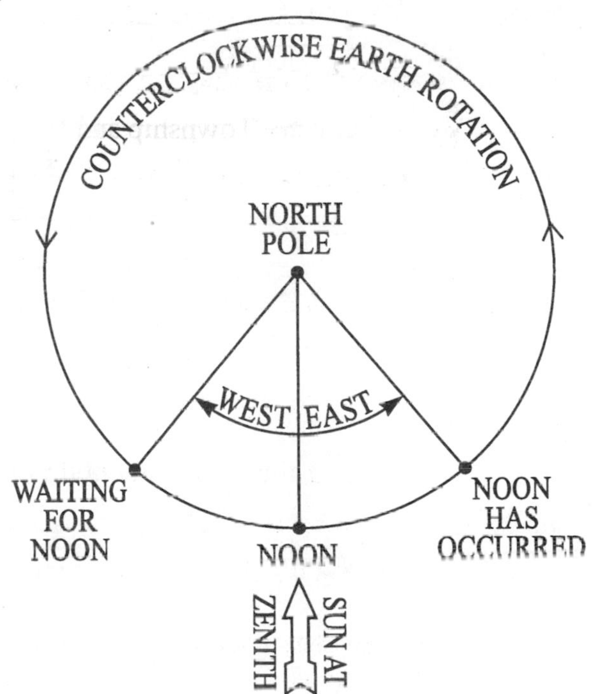

If it is solar noon at your location, what would be your longitude if the chronometer read:

1. 6:00 p.m.
2. 6:00 a.m.
3. 10:00 p.m.
4. 4:30 p.m.
5. 1:40 p.m.
6. 3:20 p.m.

2.3 Learning Activity

Have students draw a diagram, or sketch a map that illustrates the effect of the Public Lands Survey System on human affairs in the United States.

Name ______________________________ Instructor ____________________

Date ______________________________ Section Number ______________

UNIT 4 - Earth-Sun Relationships

For use with Chapter 3

4.1 Determining the Sun Angle at Noon

Reference Pages in Text: Chapter 3, pp. 78 - 89

Materials Needed:
Student Supplied: pencil
Instructor Supplied: A globe may help the student visualize the following discussion

Purpose: This exercise introduces you to the fundamental geometric relationships of Earth with the sun. These relationships determine the intensity, duration, and distribution of incoming solar radiation on our planet's surface.

A. Earth is nearly a perfect sphere. The incoming rays of the sun are parallel with one another. An infinite number of the sun's rays strike one-half of Earth's spherical surface constantly. But, because the rays are parallel and because Earth is a sphere, only one of these rays can strike Earth vertically or at a 90° angle. An observer standing at this point would see the sun directly overhead. At this same instant the sun would also be directly over the observer's meridian. When the sun is directly over a meridian, it is solar noon at that longitude.

The latitude that receives the direct ray of the sun is called the declination latitude or the sun's declination. As one proceeds north or south of the declination latitude the sun will appear at lower angles. Figure 3.18 in the textbook illustrates that the declination of the sun changes throughout the year. When the sun is directly above the Equator the declination is zero degrees. When the sun is directly above the Tropic of Cancer the declination is 23.5 degrees north. When the sun is directly above the Tropic of Capricorn the declination is 23.5 degrees south. The values of declination will always be between these values.

The highest angle that the sun appears above the horizon during a day is called the noon sun angle (and the time is solar noon). This angle can be determined for any latitude, if the declination latitude is known. First determine the distance in degrees between the observer's latitude and the declination latitude and subtract this value from 90 degrees. For example if the observer's latitude is 45 degrees north and the declination is 20 degrees north, the distance between them is 25 degrees. This subtracted from 90 degrees is 65 degrees. At noon the observer would see the sun 65 degrees above the horizon. Because the declination latitude was south of the observer's latitude, the sun would appear above the southern horizon.

If the declination latitude is north of the observer's latitude, the sun would appear above the northern horizon. This is true regardless of the hemisphere. For example if the declination latitude is 10 degrees south and the observer is 20 degrees south, the sun would appear above the northern horizon. There are situations where the sun would appear directly overhead as well as situations where the sun would appear on the horizon at noon.

The steps for determining the noon sun angle are as follows:

1. Determine the latitude of the sun's declination. This can be obtained from Figure 3.18. For the purpose of this exercise it will be given.
2. Determine the latitude of the observer. This will also be given.
3. Calculate the angular distance between these two latitudes. If the latitudes are in the same hemisphere, that is, on the same side of the Equator, they would be subtracted. If they are in different hemispheres, on opposite sides of the Equator, they would be added.
4. Subtract the answer in #3 from 90 degrees. This is the sun angle at solar noon.
5. Finally, determine if the sun would appear above the north or south horizon. It is also possible for the sun to be directly overhead when the noon sun angle is 90 degrees, or to appear on the horizon when the noon sun angle is zero degrees.

Determine the following noon sun angles and specify whether the observer would see the sun directly overhead or toward the northern or southern horizon (overhead, north, or south).

	Observer's Position	Declination Latitude	Noon Sun Angle	Sight Direction
1.	Equator	0°	______	______
2.	25°N	0°	______	______
3.	North Pole	0°	______	______
4.	20°N	20°N	______	______
5.	60°N	10°N	______	______
6.	5°	20°N	______	______
7.	40°N	10°S	______	______
8.	40°S	10°S	______	______
9.	10°S	20°S	______	______
10.	70°N	20°S	______	______

4.2 Determining Latitude

Reference Pages in Text: Chapter 3, pp. 78 - 89

Materials Needed:
Student Supplied: pencil
Instructor Supplied: A globe may help the student visualize the following discussion.

Purpose: This exercise illustrates the ability to determine the latitude of a point on Earth's surface if the noon sun angle and the declination latitude are known.

A. For many centuries, navigators have been able to determine their latitude by utilizing the dependable geometric relations that exist between Earth and sun. If the noon sun angle can be obtained it can be used to determine the angular distance that the navigator is displaced from the declination latitude. This will yield the latitude of the navigator. Remember that the sun is directly overhead at the declination latitude at solar noon. If the observer was also at this latitude the sun would obviously appear directly overhead or at the observer's zenith. In this case the noon sun angle would be 90 degrees but the zenith angle would be zero degrees. If the observer were one degree of latitude north of the declination latitude the noon sun angle would be 89 degrees and the zenith angle would be one degree. For every degree of latitude the observer is displaced from the declination latitude, the zenith angle is displaced an equal amount.

Earth's tilted axis, and the reliability of Earth's motions have enabled scientists to predict with precision the declination latitude for every day of the year. The noon sun angle is commonly obtained by using an instrument called a sextant. A person proficient in using a sextant can accurately measure the altitude of the sun above the horizon at noon. The zenith angle is found by subtracting the noon sun angle from 90 degrees. The zenith angle will indicate the distance in degrees of latitude from the declination latitude. It is also necessary to note whether the sun is north or south of the person taking the sextant shot. Now it is simply a matter of adding or subtracting the zenith angle from the declination latitude to find the observer's latitude.

The steps for determining the latitude are as follows:

1. Determine the declination latitude. This will be given.
2. Determine the noon sun angle. This would normally be done with a sextant, but for this exercise it will be given.
3. Subtract the noon sun angle from 90 degrees to find the zenith angle.
4. Note whether the sun is north or south of the observer.
5. Add or subtract the zenith angle from the declination latitude to determine the observer's latitude. If the unknown latitude is in the same hemisphere and poleward of the declination latitude the zenith angle is added. If the unknown latitude is Equatorward of the declination latitude or in the opposite hemisphere, the zenith angle and the declination latitude are subtracted from each other depending on which one has the greater value.

Here are some examples:

The declination latitude is 10 degrees north. A navigator sights the noon sun at 50 degrees south. Find the latitude. Solution: The noon sun angle is subtracted from 90 degrees to give a zenith angle of 40 degrees. Because the sun is south of the navigator it means that the unknown latitude is north of the declination latitude and therefore in the same hemisphere. In this case the zenith angle would be added to the declination latitude to give a latitude of 50 degrees north.

The declination latitude is 10 degrees north. This time the noon sun angle is 82 degrees but it is above the northern horizon. Find the latitude. Solution: The noon sun angle is subtracted from 90 degrees to give a zenith angle of 8 degrees. In this case the unknown latitude is south of the declination latitude or towards the Equator, therefore the zenith angle is subtracted from the declination latitude to give a latitude of 2 degrees north.

The declination latitude is 10 degrees south. The noon sun is 60 degrees above the southern horizon. What is the latitude? Solution: The noon sun angle is subtracted from 90 degrees to give a zenith angle of 30 degrees. This time the unknown latitude is north of the declination latitude and not only towards the Equator but past it in the opposite hemisphere. Now the zenith angle is subtracted from the declination latitude to give a latitude of 20 degrees. In this case the latitude is in the opposite or Northern Hemisphere.

Determine the latitudes in the following problems. Include N or S in your answer unless the latitude is on the Equator.

1. The crew of a schooner is sailing in tropical waters off the Galapagos Islands. The navigator uses the sextant to obtain a noon sun angle of 90°. The sun's declination is zero degrees. What is the latitude of the schooner?

2. A crew is sailing off Cape Cod. The noon sun is above the south horizon at an angle of 68°. The sun's declination on this day is 20° north. What is the latitude?

3. Somewhere south of Hawaii the navigator of a sailboat finds the noon sun above the northern horizon at 80°. The sun's declination is 20° north. What is the latitude?

4. Off the South American coast near Rio de Janeiro the navigator on board a schooner "shoots" the noon sun above the northern horizon and obtains an angle of 57°. The sun's declination is 10° north. What is the latitude of this vessel?

5. A polar explorer takes a sun reading and determines the noon sun angle to be 20° above the southern horizon. The sun's declination is 20° north. What is the position of this explorer?

6. Another polar explorer takes a sun reading and observes the noon sun 20° above the northern horizon. The sun's declination is 20° south. What is the position of this explorer?

4.3 Learning Activity

Another use of sun angles is in the planning and design of solar collectors for an active solar system for heating a house. The maximum energy is obtained when the solar collector is oriented perpendicular to the incoming solar radiation. This should be planned for the winter time. In order to achieve this orientation, the noon sun angle for the latitude of the house would need to be determined during the winter season. The declination latitude on the first day of winter in the Northern Hemisphere is approximately 23 degrees south.

Have your students determine the orientation of flat solar collectors that would achieve maximum efficiency on the first day of winter for the latitude of your location.

Name ______________________________ Instructor ______________

Date ______________________________ Section Number ______________

UNIT 5 - Temperature Controls

For use with Chapter 4

5.1 Altitude

Reference Pages in Text: Chapter 4, pp. 97, 100

Materials Needed:
Student Supplied: pencil
Instructor Supplied: none

Purpose: This exercise gives you practice in determining temperature changes related to the effect of altitude.

A. If two places were located at the same latitude but at different elevations, the one at the higher elevation would be cooler. This is assuming the other environmental factors that have an effect on air temperature are the same at both places

Many observations over the years bear out the fact that temperature drops at an average rate of 6.5°Celsius for every 1,000 meters of altitude gained. This is called the normal (environmental) lapse rate. It is important to note that this is an average. On any given day the atmospheric conditions will likely cause some variation from this average.

To determine the difference in temperature, caused by altitude, between two places, first determine the difference in elevation. Then multiply 6.5°C by the number of thousands of meters that separate the two places. This will yield the temperature difference. If the temperature of one of the places is known, it is possible to estimate the second location's temperature by subtracting or adding the calculated temperature difference from the first location's temperature. It is subtracted if the second location is at a higher elevation than the first location. It is added if the second location is at a lower elevation than the first location. The steps are:

1. Determine the elevation difference between the two locations.
2. Divide this answer by 1,000.
3. Multiply by 6.5° to find the difference in temperature.
4. Either add or subtract the temperature difference to the temperature of the known location to find the temperature of the unknown location.

Determine the following:

1. Colorado Springs, Colorado is located near the base of Pikes Peak. The elevation of the city is 1,881 meters. The top of Pikes Peak is 4,301 meters. If average atmospheric conditions prevail, how much temperature drop should a hiker expect to encounter because of the elevation differential between Colorado Springs and the top of Pikes Peak?

2. If the temperature in Colorado Springs was 5°C, what is the temperature at the top of Pikes Peak? Consider only the effects of the environmental lapse rate.

3. The temperature at the top of Pikes Peak is –5°C. What is the temperature in Colorado Springs caused by the elevation difference?

4. Your automobile's radiator is protected by anti-freeze to –30°C. Your route includes going over Independence Pass in Colorado which has an elevation of 3,687 meters. Before going over the pass, you stop in Salida, Colorado for fuel and the temperature there is –22°C. The elevation of Salida is 2,280 meters. Determine the temperature at the pass.

   Do you think you should add some more anti-freeze to your radiator?

5.2 Land-Water Distribution

Reference Pages in Text: Chapter 4, p. 98

Materials Needed:
Student Supplied: red and blue pencil
Instructor Supplied: none

Purpose: This exercise will give you an understanding of the basic characteristics of locations under a maritime influence and continental influence.

The sun's energy will strike equally at two places at the same latitude assuming the atmospheric conditions are the same. If one of these locations is near a large body of water and the other is hundreds of miles inland, there will likely be a seasonal contrast in temperature between these two locations.

A. San Francisco, California (37° 45'N) and St. Louis, Missouri (38° 40'N) are nearly the same latitude but are far different in longitude and in environmental setting. The following are the average monthly temperature for the two cities:

Month	San Francisco Temperature (°C)	St. Louis Temperature (°C)
January	9.2	–0.1
February	10.5	1.5
March	11.8	5.9
April	13.2	12.7
May	14.6	17.9
June	16.3	23.4
July	17.1	25.6
August	17.1	24.9
September	17.7	20.8
October	15.8	14.7
November	12.7	6.7
December	10.1	1.6

Plot the above data on the graph which follows. Use a blue pencil for the San Francisco data and a red pencil for the St. Louis data.

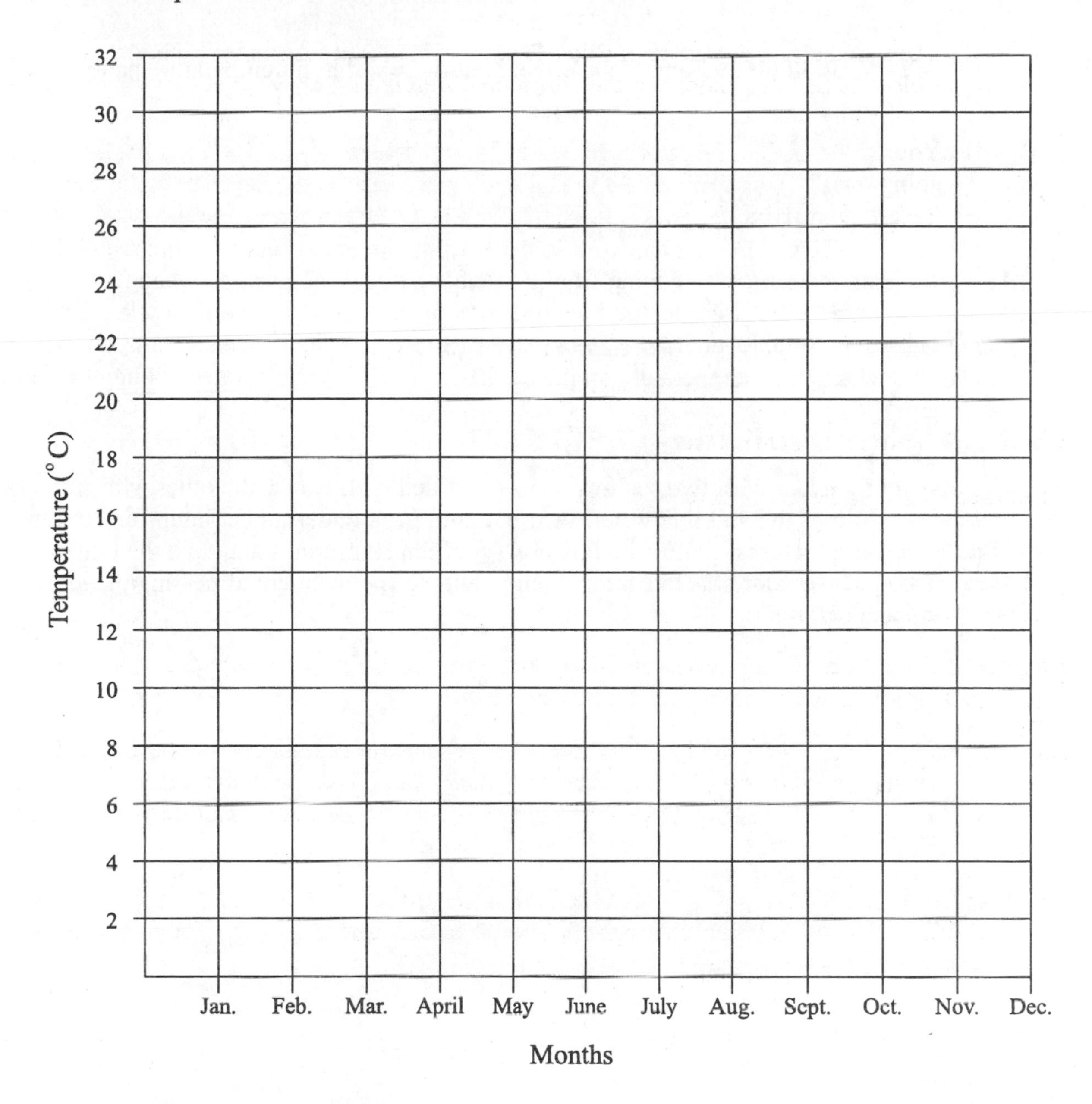

B. Answer the following questions:

1. The average annual temperature is calculated by adding the twelve monthly averages and dividing by 12. What is the average annual temperature of St. Louis?
   San Francisco?

2. The annual range of temperature is calculated by subtracting the lowest average monthly temperature from the highest average monthly temperature. What is the annual range of temperature for St. Louis?

   San Francisco?

3. In what seasons do the maximum differences in temperature occur between these two cities.

   a.

   b. r

4. Why do St. Louis and San Francisco exhibit these different temperature patterns?

5.3 Learning Activities

A. Have students record, or determine from a reliable source, the periodic changes (every 3-6 hours) in temperature for the local community over a period of one week. List these temperatures in a table or display them in graph form. Have the students analyze and explain the changes in temperature on the basis of what they have learned about atmospheric heating.

B. Provide your student with two shallow pans–one filled with water, the other with soil. Take the temperature of the soil and water, then put both pans under a heat lamp for 10 minutes. Retake the temperatures. Allow both pans to cool for 10 minutes and take the temperatures once again. Have your students relate their results to the differential heating of our oceans and continents.

Name ______________________________ Instructor ______________

Date ______________________________ Section Number ______________

UNIT 6 - Atmospheric Pressure

For use with Chapter 5

6.1 Global Variations in Horizontal Pressure Distribution

Reference pages in Text: Chapter 5, pp. 119 – 120, 129 – 132

Materials Needed:
Student Supplied: blue and red pencils
Instructor Supplied: Atlas (optional)

Purpose: This exercise illustrates the latitudinal and longitudinal variation in surface pressure over Earth's surface; the seasonal variation in surface pressure; and role of land and water distribution in surface pressure variations.

A. Using the average sea level pressure maps provided in the textbook, or found in most atlases, do the following:

1. First study the latitudinal (north-south) variation in surface pressure that occurs over Earth's surface. Then on the graph below plot the observed January surface pressure along the 40° W meridian of longitude for the latitudes indicated. Finally, construct a latitudinal profile curve by connecting the plotted points (use a blue pencil). On the same graph repeat the above steps for July (use a red pencil).

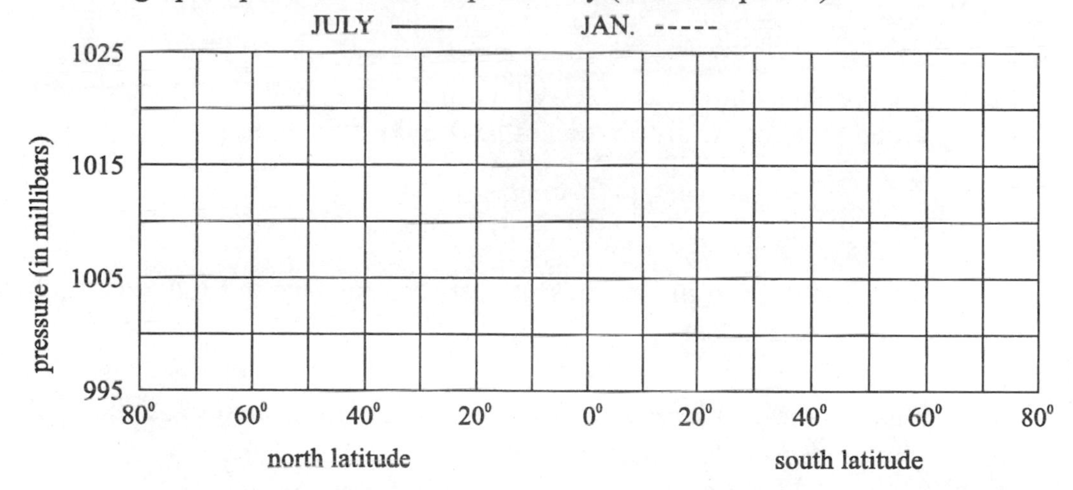

a) On the January and July profiles you just drew, label the following: Equatorial Low (EL), Subtropical Highs (STH), and Subpolar Lows (SPL).

b) Describe the seasonal latitudinal shift that these pressure belts undergo. Why do they shift?

c) Describe the seasonal variation in magnitude that occurs within these pressure belts. Why do these variations in magnitude occur?

2. First study the longitudinal (east-west) variation in surface pressure that occurs at various latitudes over Earth's surface. Then on the graph below plot the observed January surface pressure along the 40° N parallel of latitude for the longitudes indicated. Finally, construct a longitudinal profile curve by connecting the plotted points (use a blue pencil). On the same graph repeat the above steps for July (use a red pencil).

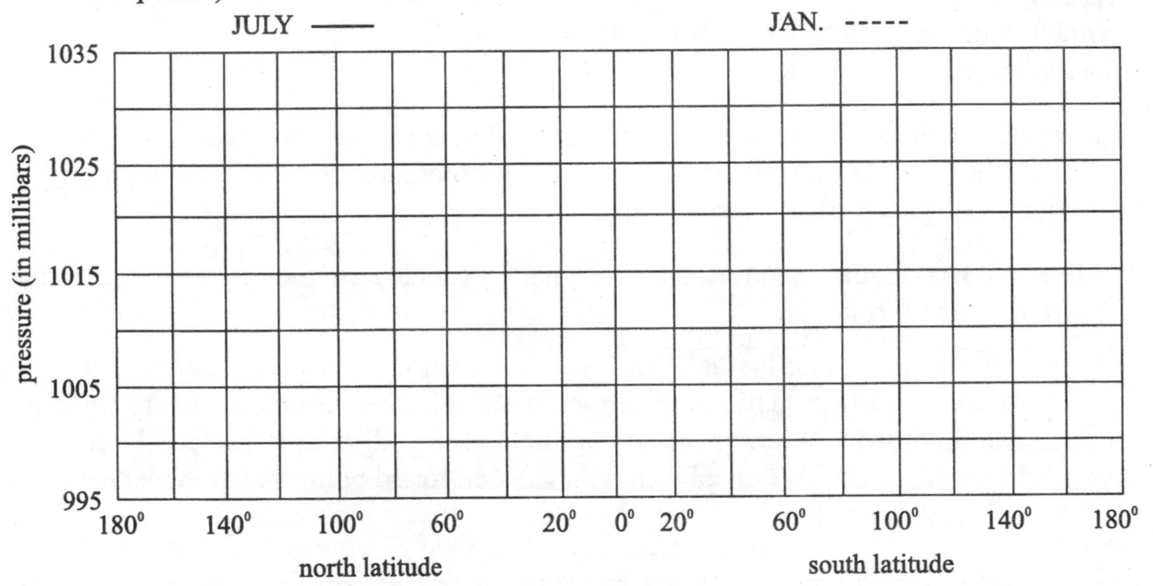

a) The pressure variations observed during January and July along the 40th parallel of latitude reflect the distribution of land and water. Why?

b) Is the variation in surface pressure between January and July greatest over North America or Eurasia?

c) What do these observations indicate? *Larger landmasses heat and cool to a greater degree than smaller ones.*

3. Once again, look at the January and July maps of average sea level pressure. Answer the following questions:

a) Why is the subtropical high pressure belt more continuous (i.e. linear not cellular) in the southern hemisphere than in the northern hemisphere in July?

b) During July, what area of the United States exhibits the lowest average pressure?
Why?

c) Pressure and temperature appear to have a significant negative correlation (i.e. as temperature decreases, pressure increases). Therefore based upon the January pressure map, where would you expect to find the coldest temperatures in the world?

d) Look at Figure 4.19—Average January Temperatures—on page 112. Does the relationship appear valid?
Why?

6.2 Air Pressure Demonstration

Reference Pages in Text: Chapter 5, pp. 117 - 118

Materials Needed:
Student Supplied: none
Instructor Supplied: 1 gallon metal can with lid, hot plate, water, and glove

Purpose: Since the pressure within our bodies equalizes the pressure exerted by the atmosphere we are unaware of the pressure associated with the weight of the atmosphere. The following experiment will be performed by your instructor to demonstrate atmospheric pressure.

A. Perform the following steps:

1. Add a small amount of water (1 cup) to the clean 1 gallon can.
2. With the lid off heat the can on the hot plate until the water boils—let boil for several minutes.
3. Using your gloved hand remove the can from the burner and secure the lid tightly. Let the can cool—preferably near an open window if it is cold outside.
4. Soon the can will begin to collapse. Why?

6.3 Computation Problems

Reference Pages in Text: Chapter 5, pp. 117 - 118

Materials Needed:

Student Supplied: Pencil, calculator (optional)
Instructor Supplied: none

Purpose: This exercise will give you experience in converting pressure units and in calculating changes in vertical pressure.

A. Make the following pressure conversions:

1018 mb = _____ mm = _____ inches

_____ mb = 750 mm = _____ inches

_____ mb = _____ mm = 29.8 inches

B. Atmospheric pressure decreases at the rate of 0.036 mb per foot as one ascends through the lower portion of the atmosphere. The Sears Tower in Chicago, Illinois once the world's tallest building at 1450 ft. If the street level pressure is 1020.4 mb, what is the pressure at the top of the Sears Tower?

C. If the difference in atmospheric pressure between the top and ground floor of an office building is 13.5 mb, how tall is the building?

D. A single story of a building is 12 feet. You enter an elevator on the top floor of a building and wish to descend five floors. The elevator has no floor markings—only a barometer! If the initial reading was 1003.2, at what pressure reading would you get off?

6.4 Learning Activities

A. Have your students keep a daily record of atmospheric pressure and temperature. After obtaining two weeks of records have your students discuss the relationship of these two variables. Have them graph the results.

B. Obtain pressure observations for two typical weeks during the summer season. Have your students note the variability in pressure during the two selected weeks. Also have them note the highest and lowest pressures recorded during both weeks. Have them explain why "highs" are higher during the winter and "lows" are lower during the summer.

Name ______________________________ Instructor ________________

Date ______________________________ Section Number ____________

UNIT 7 - WIND

For use with Chapter 5

7.1 Wind Components and Flow

Reference Pages in Text: Chapter 5, pp. 120 - 123

Materials Needed:
Student Supplied: pencil
Instructor Supplied: none

Purpose: This exercise examines the interaction among pressure gradient, Coriolis effect, and friction.

A. Figure 5.5 in the text illustrates a balanced geostrophic and surface wind system for the Northern Hemisphere.

1. What is the most significant difference between the two wind systems?

2. Why does friction cause a surface wind to have a different direction and magnitude than an upper air wind?

3. On the diagram below draw a balanced wind system for a geostrophic and surface wind in the Southern Hemisphere.

Geostrophic Wind

High pressure

Low pressure

Surface Wind

High pressure

Low pressure

7.2 The Global Wind System and Ocean Currents

Reference Pages in Text: Chapter 5, pp. 133 - 142

Materials Needed:
Student Supplied: blue and red pencils
Instructor Supplied: none

Purpose: This exercise examines the global wind system and its influence on the world pattern of ocean currents.

A. Using Figure 5.23 as a guide, draw the major ocean currents on the outline map below. Use a blue pencil for cold currents and a red pencil for warm currents.

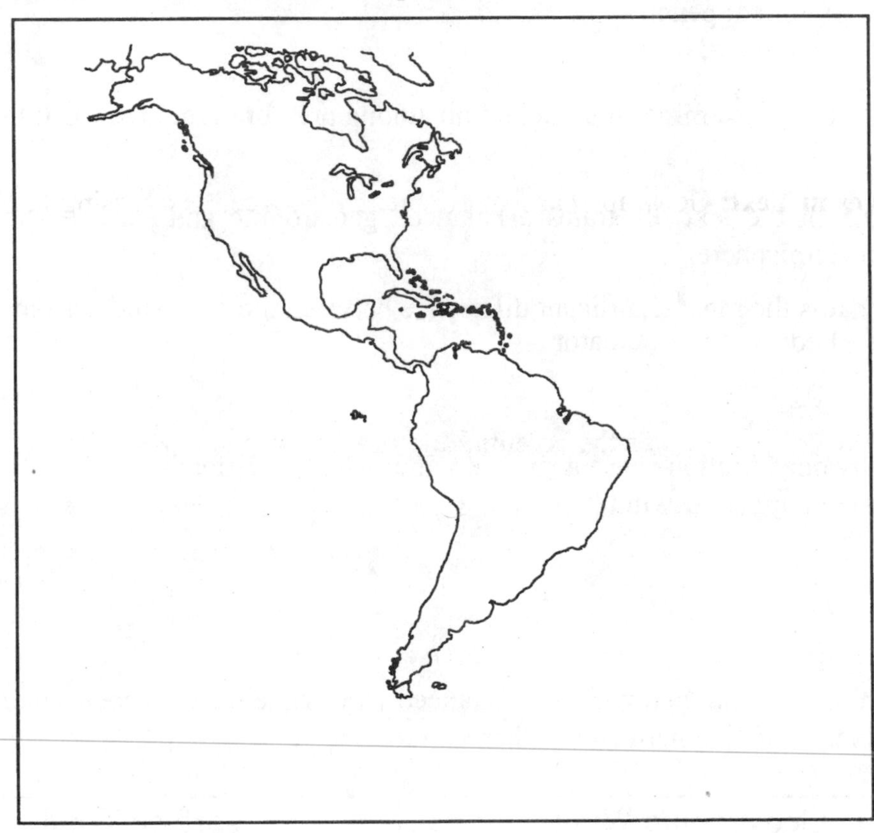

B. Carefully review Figure 5.13b (Average July Sea Level Pressure) and Figure 5.14 (Earth's Pressure and Wind Systems). Using a regular lead pencil use broad arrows to illustrate the global winds over the ocean areas depicted above.

1. What is the relationship between Earth's wind system and ocean currents?

2. If you wished to sail from England to New York City, what route would you take? Why?

3. If Earth were to rotate clockwise instead of counterclockwise, what impact would that have on our surface winds?

   How would the pattern of ocean currents change?

7.3 Wind Power

Reference Pages in Text: Geography as Environmental Science: Harnessing the Wind, Chapter 5, pp. 136 - 137

Materials Needed:
Student Supplied: pencil, calculator
Instructor Supplied: none

Purpose: This exercise illustrates the potential use of wind as a source of power.

A. The amount of power which can be generated by wind is determined by the equation:

$$P = \tfrac{1}{2}D \times S^3$$

where P is the power in watts, D is the density, and S is the wind speed in meters/sec. Since $D = 1.293$ kg/m^3 we can rewrite the equation as:

$$P = 0.65 \times S^3$$

1. How much power is generated by the following wind speeds?

   2 m/s = _____ watts   6 m/s = _____ watts

   4 m/s = _____ watts   8 m/s = _____ watts

2. Describe the increase in power which occurs when the wind speed is doubled.

3. Since wind power increases significantly with increased wind speed, very windy locations are ideal locations for "wind farms." Cities A and B both have average wind speeds of 6 meters/second. However, City A tends to have very consistent winds, while City B tends to have half of its winds at 2 meters/second and the other half at 10 meters/second. Which site would be the best location for a wind generation plant? Show your work.

   City A =

   City B =

7.4 Learning Activities

A. Without reference to the textbook have students draw a diagram, or a series of diagrams, that illustrates one of the following: Earth's surface wind belts; monsoons; land and sea breezes; mountain and valley breezes; chinook winds.

B. Lead a discussion with your students concerning the associations among Earth-sun relationships; insolation; the horizontal distribution of temperature on Earth; and the horizontal distribution of pressure and winds on Earth.

C. Provide data so that the class can keep an accurate log of wind speed and direction for the local community for a minimum of two weeks. They should also keep a log of temperatures and precipitation during that same period. Have students discuss the relationship between wind direction, wind speed, and the observed weather.

Name ______________________________ Instructor ______________

Date ______________________________ Section Number ______________

UNIT 9 - Stability and Instability

For use with Chapter 6

9.1 Determining Stability Conditions - Constant Environmental Lapse Rate

Reference Pages in Text: Chapter 6, pp. 160 - 162

Materials Needed:

Student Supplied: red and blue pencil, straight edge
Instructor Supplied: none

Purpose: This exercise illustrates the fundamental relationship of the adiabatic lapse rate and environmental lapse rate relative to the concept of stability/instability.

A. Surface observations indicate that a parcel of air has an air temperature of 27°C and a dew point temperature of 11°C. First calculate the Lifting Condensation Level - LCL (review Unit 8) and then on the two graphs below plot the adiabatic temperature profile which would result if the parcel were lifted to 6000 meters (use a blue pencil). Recall the Dry Adiabatic Lapse Rate = 10°C/1000 m while the Wet Adiabatic Lapse Rate = 5°C/1000 m.

Superimpose the Environmental Lapse Rates (the vertical temperature profile) in red pencil. On the left hand graph (Day 1) assume the ELR = 3°C/1000 m; on the right hand graph (Day 2) assume the ELR = 12°C/1000 m.

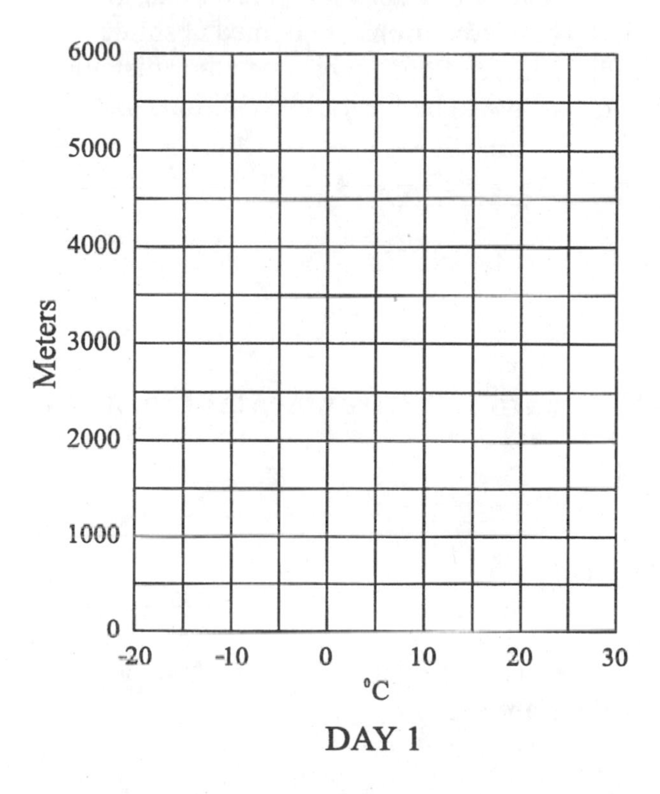

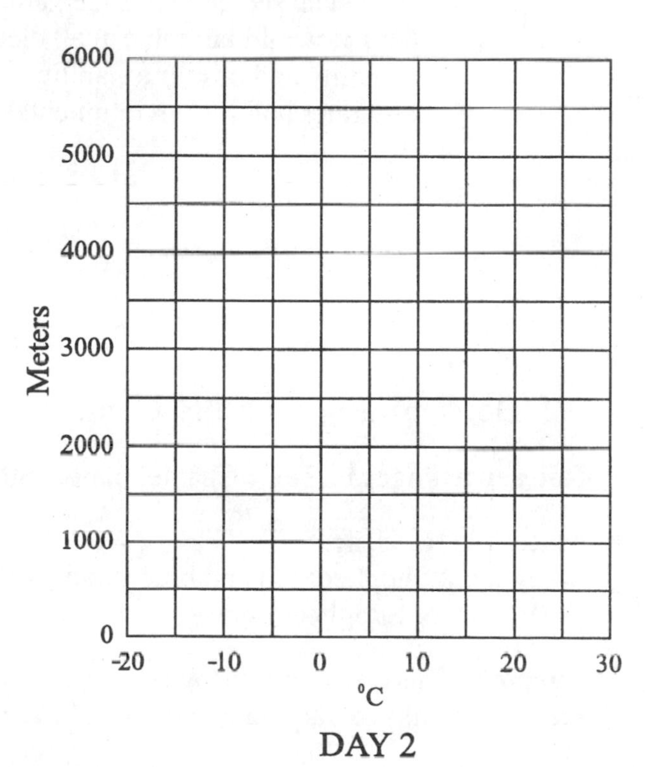

Next determine if the parcel of air would be stable, unstable, or neutral if it did rise through the atmosphere. Determining this involves nothing more than asking the question: If the parcel were lifted to a specific elevation, cooling at the adiabatic rates–blue line, would it be warmer (and therefore *unstable*), colder (and therefore *stable*) or the same temperature (and therefore *neutral*) as the environmental air–redline–at that level. Stable air will sink and unstable air will rise until they reach a layer of air with the same temperature.

Answer the following questions:

1. If the parcel of air were lifted to 3000 meters would it be stable or unstable on:
   Day 1?
   Day 2?

2. If the parcel of air were lifted to 5000 meters would it be stable or unstable on
   Day 1?
   Day 2? *e*

3. What type of clouds (stratiform or cumuloform) would you expect to develop on:
   Day 1? Why?

   Day 2? Why?

4. As you can see on Day 2 the parcel of air would be unstable at all elevations while on Day 1 it would be stable at all elevations. These conditions are termed absolute instability and absolute stability. What does the relationship between the adiabatic lapse rates and the environmental lapse rates have to be for these conditions to exist?

9.2 Determining Stability Conditions - Fluctuating Environmental Lapse Rate

Reference Pages in Text: Chapter 6, pp. 160 - 162

Materials Needed:
Student Supplied: red and blue pencil, straight edge
Instructor Supplied: none

Purpose: This exercise illustrates how a fluctuating environmental lapse rate, which is more realistic, results in varying conditions of stability and instability.

A. Conditions of absolute stability and instability seldom exist since the environmental lapse rate usually varies far more than our two examples above. The example below is far more realistic.

Once again surface observations indicate that the air temperature is 27°C and the dew point temperature is 11°C. However, radiosonde reports reveal that the upper air temperatures are as follows:

6000 m - (–10°C)	4000 m - (–7°C)	2000 m - (10°C)	0 m - (27°C)
5500 m - (–5°C)	3500 m - (–5°C)	1500 m - (12°C)	
5000 m - (0°C)	3000 m - (–2°C)	1000 m - (23°C)	
4500 m - (–4°C)	2500 m - (9°C)	500 m - (24°C)	

1. On the graph below once again plot the adiabatic lapse rate in blue and the observed upper-air temperatures (environmental lapse rate) in red pencil.

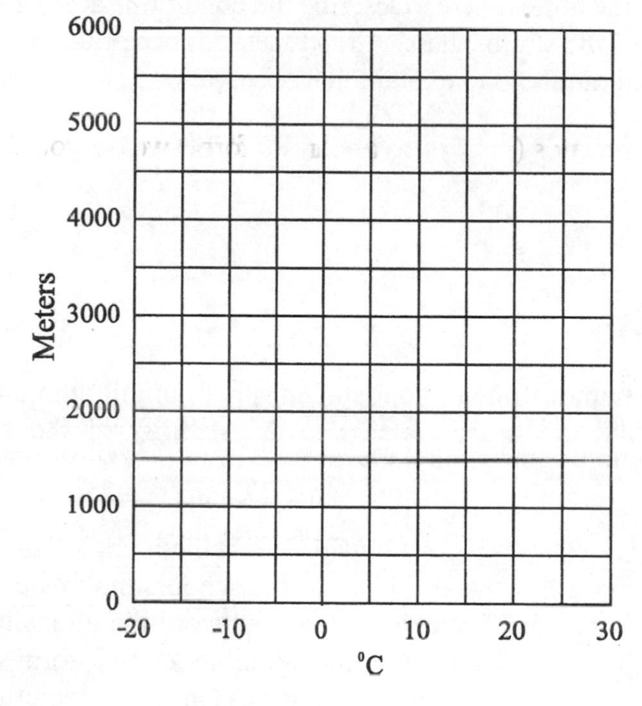

2. If you plotted the information correctly the lines cross at 0 meters, 1500 meters, 2700 meters, and 4300 meters. What does it mean when the two lines cross?

   The stability is neutral - a parcel will neither rise not sink.

3. Next determine if the parcel of air would be stable or unstable between:

   0–1500 meters?

   1500–2700 meters?

   2700–4300 meters?

   4300–6000 meters?

4. Answer the following (be sure to remember the height of the LCL in answering): Assume that the parcel was lifted to the elevations indicated but once it reached that height the lifting mechanism dissipated. Describe the response of the parcel.
   a) lifted to 1200 meters

   b) lifted to 2500 meters

   c) lifted to 3000 meters

5. Look at Figure 4.3 (page 96) which illustrates the average vertical temperature structure of the atmosphere. Describe the conditions at the top of the troposphere and the bottom of the stratosphere. Clouds seldom penetrate into the stratosphere. How does this diagram help to explain that observation?

9.3 Learning Activity

A. This activity will demonstrate to your students that rainfall amounts, as well as intensity and duration of rainfall, are highly related to the stability of the atmosphere which can be assessed by cloud form.

Have your students establish a reasonably dense network of rain gauges (3 pound coffee cans will do) within a three-mile radius of the school. Each student should be responsible for one or two rain gauges. Have your students record the intensity, duration, and amount of rainfall after the passage of a storm which exhibits cumuloform clouds and again after the passage of a storm exhibiting stratiform clouds. On maps depicting the study area enter the rainfall amounts and draw isohyets. Have the students compare the rainfall characteristics (including the spatial variability) associated with these different stability conditions.

Name ______________________________ Instructor ____________________

Date ______________________________ Section Number ______________

UNIT 11 - Surface and Upper Air Map Analysis

For use with Chapter 7

11.1 Weather Map Analysis

Reference Pages in Text: Chapter 7, pp. 186 - 189

Materials Needed:

Student Supplied: blue and red pencils
Instructor Supplied: Atlas (optional)

Purpose: This exercise will help you to better understand the weather associated with a middle-latitude cyclone and to forecast the weather most likely to occur as the cyclone moves across the United States.

A. Describe the weather associated with:

1. A warm front

2. A cold front

B. The figure below depicts a middle-latitude cyclone.

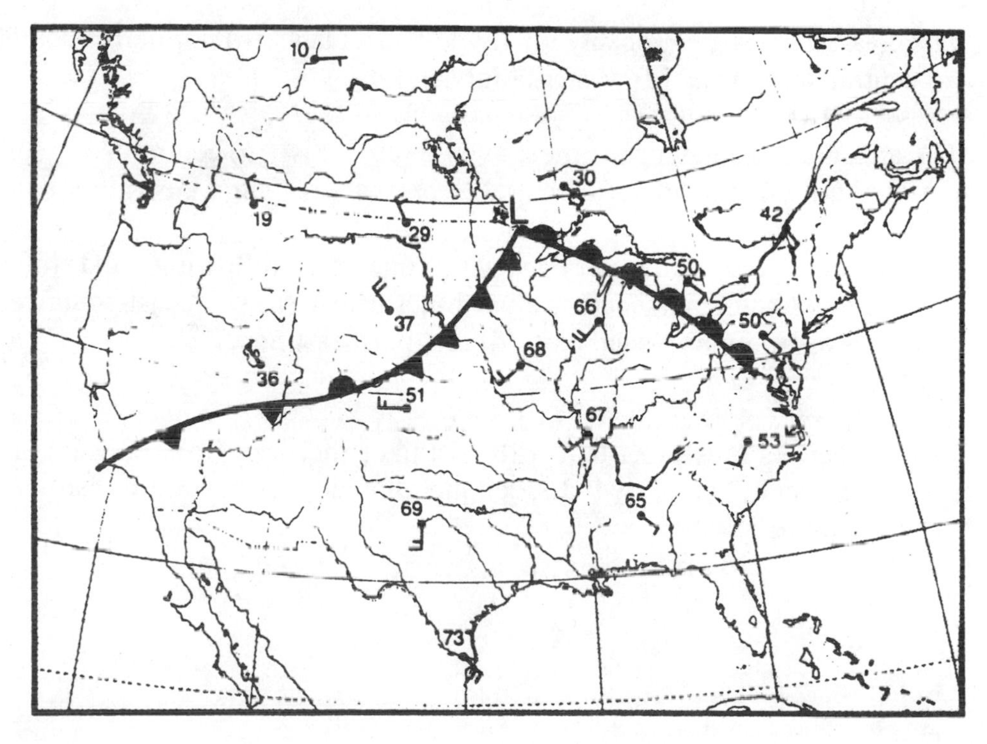

Assume the storm system is traveling due east at 30 miles per hour. Answer the following questions:

1. The wind in central Indiana is most likely from the (a) east (b) northwest (c) southeast (d) west

2. Which of the following locations should experience the greatest total precipitation in the next 36 hours? (a) Toronto (b) Chicago (c) Rapid City, South Dakota

3. The location most likely to experience thunderstorms in the next 12 hours is: (a) northwest Iowa (b) southeast Montana (c) northern Wisconsin (d) northern New York

4. During the next 12 hours Albany, New York should experience which of the following: (a) thunderstorms followed by cool weather and low humidity; (b) steady rainfall followed by warmer weather and higher humidity; (c) clear skies followed by showers and falling temperatures; (d) severe weather followed by clearing and cooler temperatures.

5. When will the center of low pressure most likely reach the Atlantic coast: (a) 12 hours (b) 24 hours (c) 36 hours (d) 48 hours

6. Which location is most likely to receive snow during the next 12 hours: (a) Duluth, Minn. (b) Billings, Mont. (c) Buffalo, NY (d) Washington D.C.

7. The location with the highest barometric pressure would be (a) central Georgia; (b) southern Maine (c) northern Nevada (d) central Montana

8. The clearest skies are probably in: (a) northern Idaho (b) northern Michigan (c) central Wisconsin (d) western Maryland

9. The area probably having the most intense rainfall is: (a) central New York (b) southern Illinois (c) north-central Nebraska (d) central Georgia

10. Buffalo, New York will experience which one of the following wind shifts during the next 36 to 48 hours: (a) southeast-north-northwest (b) southeast-southwest-northwest (c) northwest-northeast-southeast (d) east-northeast-north

D. Describe the changes in weather that will most likely occur at Boston, Mass during the next 36 hours. Specifically discuss wind speed and direction, temperature, barometric pressure, cloud cover, and precipitation.

11.2 Upper Air Analysis

Reference Pages in Text: Chapter 7, pp. 191 - 194

Materials Needed

Student Supplied: pencil and eraser
Instructor Supplied: atlas (optional)

Purpose: This exercise will help you understand the role of the upper air winds in steering surface storm systems as well as their role in the strengthening or dissipation of surface storms.

The figure below depicts the upper air wind flow (30,000') over North America.

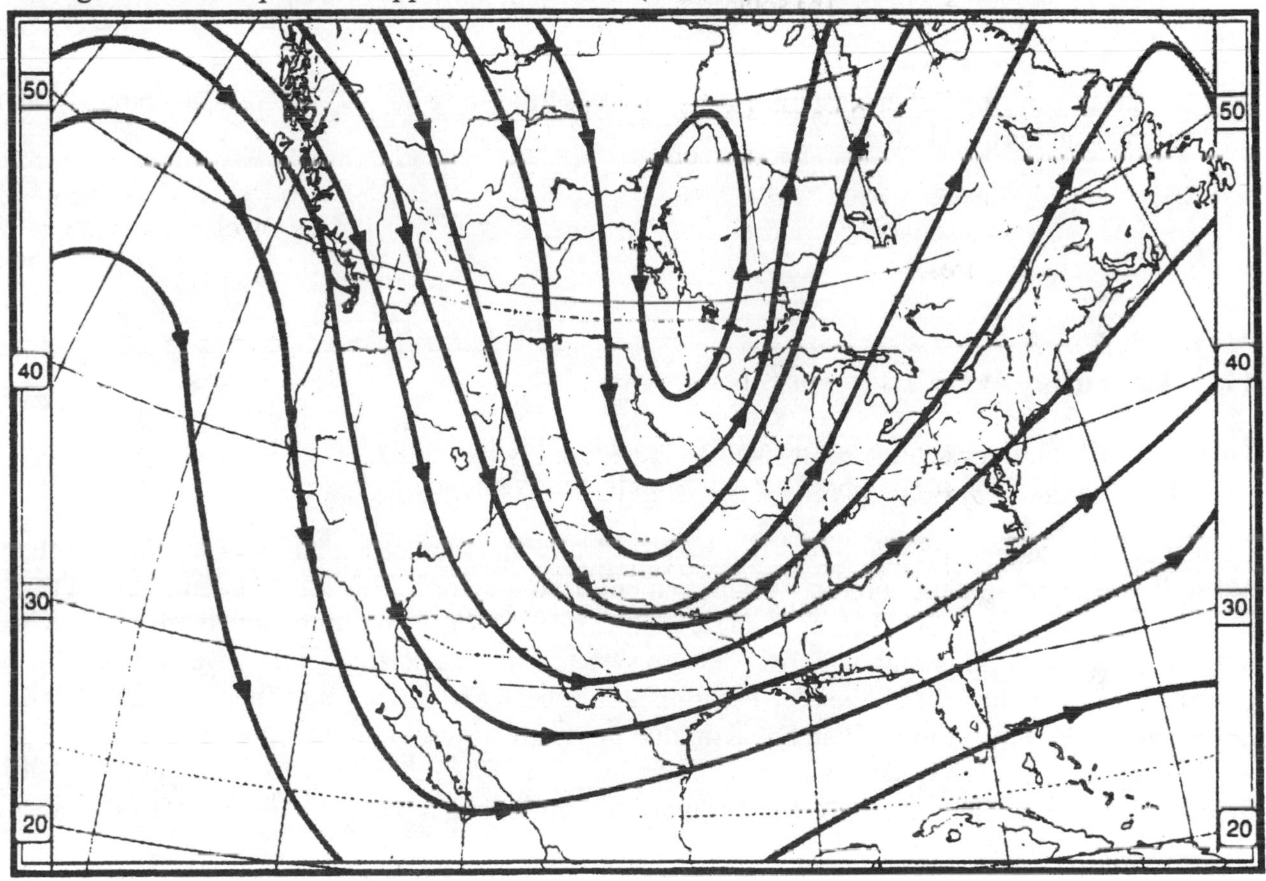

A. On the above figure label the two upper air ridges (R), and the upper air trough (T).

B. As the air in the upper atmosphere flows from an upper air ridge to an upper air trough it undergoes convergence. What impact does that have on surface storms in that region?

C. As the air in the upper atmosphere flows from an upper air trough to an upper air ridge it undergoes divergence. What impact does that have on surface storms in that region?

D. What region of the United States is least likely to experience severe thunderstorms with this upper air pattern?
most likely?

E. You are the meteorologist in Cleveland, Ohio. Would you expect temperatures in Cleveland to be normal, below normal, or above normal under this upper air pattern?

Why?

F. A storm develops over eastern Colorado. Describe its path and changes in intensity as it travels across the United States.

11.3 Learning Activity

This activity will illustrate the predictability (or lack of predictability) of surface weather movement and the role of the upper air winds on that movement.

Collect a 5-day sequence of daily weather maps. Ideally this sequence should be several months old so that students cannot "predict" the movement from memory. Have the students study the surface maps for days 1-3. Then on a single outline map of North America have them plot the movement of the various high and low pressure systems, and associated fronts, over that three-day period. From this they should be able to estimate the speed and direction of the upper air flow during that three-day period. Then based on that upper air flow and the location of surface weather systems on day 3, have them project the location of those systems on days 4 and 5. After they are done show them the daily weather maps for days 4 and 5 and have them evaluate their predictions.

Name ____________________ Instructor ____________________

Date ____________________ Section Number ____________________

UNIT 13 - Low Latitude and Arid Climates

For use with Chapter 9 and Appendix C

13.1 Köppen Classification System

Reference Pages in Text: Appendix C

Materials Needed:

Student Supplied: pencil
Instructor Supplied: atlas

Purpose: This exercise will give you additional practice in applying the Köppen classification system to climatic data.

A. Determine the climate type of the following locations:

1.	J	F	M	A	M	J	J	A	S	O	N	D	Yr.
Temp (°C)	27	26	27	27	27	27	27	27	27	27	27	27	27
Precip (cm.)	31.8	35.8	35.8	32.0	25.9	17.0	15.0	11.2	8.9	8.4	6.6	15.5	243.8

Climate Type ____

2.	J	F	M	A	M	J	J	A	S	O	N	D	Yr.
Temp (°C)	27	26	27	27	27	27	27	27	27	27	27	27	27
Precip (cm.)	31.8	35.8	35.8	32.0	25.9	17.0	15.0	11.2	8.9	8.4	6.6	15.5	243.8

Climate Type ____

3.	J	F	M	A	M	J	J	A	S	O	N	D	Yr.
Temp (°C)	19	20	21	23	26	27	28	28	27	26	22	20	24
Precip (cm.)	5.1	4.8	5.8	9.9	16.3	18.0	17.0	17.0	24.0	20.0	7.1	3.0	149.0

Climate Type ____

4.	J	F	M	A	M	J	J	A	S	O	N	D	Yr.
Temp (°C)	13	14	17	19	22	24	26	26	26	24	19	15	20
Precip (cm.)	6.6	4.1	2.0	0.5	0.3	0	0	0	0.3	1.8	4.6	6.6	26.7

Climate Type ____

5.	J	F	M	A	M	J	J	A	S	O	N	D	Yr.
Temp (°C)	2	4	8	13	18	24	26	24	21	14	7	3	14
Precip (cm.)	1.0	1.0	1.3	1.3	2.0	1.5	3.0	3.3	2.3	2.0	1.0	1.3	20.6

Climate Type ____

6.	J	F	M	A	M	J	J	A	S	O	N	D	Yr.
Temp (°C)	22	21	21	20	17	14	14	16	20	22	22	22	20
Precip (cm.)	14.2	10.9	8.4	1.8	1.0	0.3	0	0	0.5	2.0	8.1	12.2	59.4

Climate Type ____

B. The six locations represented by the above data are listed below although not in the order of presentation. Use an atlas and your knowledge of climates to correctly match the climatic types with the proper locations.

Location	Station Data Number (1-6)
Albuquerque, New Mexico	____
Belém, Brazil	____
Benghazi, Libya	____
Bulawayo, Zimbabwe	____
Kano, Nigeria	____
Miami, Florida	____

13.2 Climatic Controls

Reference Pages in Text: Chapter 9, p. 244 (Table 9.2), p. 255 (Table 9.3)

Materials Needed:
Student Supplied: pencil
Instructor Supplied: atlas

Purpose: This exercise will help you better understand the climatic controls which influence the low latitude and arid climates.

A. Based upon skeletal remains and other scientific evidence many believe that our early ancestors had their origins about one or two million years ago in equatorial Africa and on the island of Java. It appears that the climatic patterns during that period of development were very similar to those observed today.

   1. What is the Köppen climatic type of those regions?

   2. Why do you think such a climatic type would be conducive to the early development of the human race?

B. Why does Belém, Brazil have rainfall throughout the year while Miami, Florida has a distinct wet and dry season?

C. Benghazi, Libya and Albuquerque, New Mexico both exhibit steppe climates, yet the cause (or control) of their dry conditions is quite different. Describe the primary factor responsible for the dry conditions at each location.

1. Benghazi, Libya

2. Albuquerque, New Mexico

D. Tropical B climates are located more poleward than A climates yet their daytime high temperatures are often higher. Why?

E. Kano, Nigeria has copious amounts of rainfall during the summer season. There are two sources, or causes, of this rainfall. What are they?

F. Benghazi, Libya and Bulawayo, Zimbabwe both exhibit low latitude steppe climates. However, one has a summer rainfall maximum, the other a winter rainfall maximum. Why?

13.3 Learning Activities

A. Have your students explain why the tropical (A) climates extend farther poleward in the interior of South America than they do in Africa.

B. As you move northward from the equator along the west side of a continent the following climatic regions are encountered: Af, Aw, BS and BW. Have your students discuss the climate controls which cause this sequence of climates.

Name ______________________________ Instructor ____________________

Date ______________________________ Section Number ________________

UNIT 14 - Middle-latitude and High Latitude Climates

For use with Chapter 10 and Appendix C

14.1 Climate Classification

Reference Pages in Text: Appendix C

Materials Needed:
Student Supplied: pencil
Instructor Supplied: atlas

Purpose: This exercise will give you additional practice in applying the Köppen classification system to climatic data.

A. Determine the climate type of the following locations:

1.	J	F	M	A	M	J	J	A	S	O	N	D	Yr.
Temp (°C)	–42	–47	–40	–31	–20	–15	–11	–18	–22	–36	–43	–39	–30
Precip (cm.)	0.3	0.3	0.5	0.3	0.5	0.8	2.0	1.8	0.8	0.3	0.8	0.5	8.6

Climate Type ____

2.	J	F	M	A	M	J	J	A	S	O	N	D	Yr.
Temp (°C)	11	12	13	16	17	20	22	22	21	18	14	11	17
Precip (cm.)	10.9	7.6	10.7	5.3	4.3	1.5	0.3	0.5	3.3	6.1	9.4	10.4	70.4

Climate Type ____

3.	J	F	M	A	M	J	J	A	S	O	N	D	Yr.
Temp (°C)	23	23	22	19	16	14	13	13	14	17	19	22	18
Precip (cm.)	0.8	1.0	2.0	4.3	13.0	18.0	17.0	14.5	8.6	5.6	2.0	1.3	88.1

Climate Type ____

4.	J	F	M	A	M	J	J	A	S	O	N	D	Yr.
Temp (°C)	–27	–28	–26	–18	–8	1	4	3	–1	–8	–18	–24	–12
Precip (cm.)	0.5	0.5	0.3	0.3	0.3	1.0	2.0	2.3	1.5	1.3	0.5	0.5	10.9

Climate Type ____

5.	J	F	M	A	M	J	J	A	S	O	N	D	Yr.
Temp (°C)	–4	–2	5	14	20	24	26	25	20	13	3	–2	12
Precip (cm.)	0.5	0.5	0.8	1.8	3.6	7.9	24.4	14.2	5.8	1.5	1.0	0.3	62.2

Climate Type ____

6.

	J	F	M	A	M	J	J	A	S	O	N	D	Yr.
Temp (°C)	9	9	9	10	12	13	14	14	14	12	11	9	9
Precip (cm.)	17.0	14.8	13.3	6.8	5.5	1.9	0.3	0.3	1.6	8.1	11.7	17	97.6

Climate Type ____

7.

	J	F	M	A	M	J	J	A	S	O	N	D	Yr.
Temp (°C)	–3	–2	2	9	16	21	24	23	19	13	4	–2	11
Precip (cm.)	4.8	4.1	6.9	7.6	9.4	10.4	8.6	8.1	6.9	7.1	5.6	4.8	84.3

Climate Type ____

8.

	J	F	M	A	M	J	J	A	S	O	N	D	Yr.
Temp (°C)	0	0	4	9	16	21	24	23	20	14	8	2	12
Precip (cm.)	8.1	7.4	10.7	8.9	9.4	8.6	10.2	12.7	10.7	8.1	8.9	8.1	111.5

Climate Type ____

B. The eight locations represented by the above data are listed below, although not in the order of presentation. Use an atlas and your knowledge of climates to correctly match the climatic types with the proper locations.

Location	Station Data Number (1-8)
Beijing, China	____
Point Barrow, Alaska	____
Chicago, Illinois	____
Eismitte, Greenland	____
Eureka, California	____
Lisbon, Portugal	____
New York, New York	____
Perth, Australia	____

14.2 Climatic Controls

Reference Pages in Text: Chapter 10, p. 265 (Table 10.1), p, 278 (Table 10.2), p. 288 (Table 10.3)

Materials Needed:
Student Supplied: pencil
Instructor Supplied: atlas

Purpose: This exercise will help you better understand the climatic controls which influence the middle and high latitude climates.

A. Eureka, Chicago, and New York are located within a few degrees latitude of one another, yet they represent three distinctly different climatic types. Discuss these differences and identify the primary cause, or source, of these differences.

B. The precipitation recorded at Albuquerque, New Mexico (see Unit 13) is almost twice that recorded at Point Barrow, Alaska. Yet, Albuquerque is considered a dry climate and Barrow a humid climate. Why?

C. How does Figure 6.3 (p. 153) help to explain why Eismitte, Greenland has so little precipitation?

D. What differentiates the Dw climate from the other D climates?

Why is the Dw climate type only found in Asia?

E. Why don't D climates exist in the Southern Hemisphere?

F. Csa climatic regions and Cfa climatic regions are both under the influence of subtropical high pressure cells during the summer, yet Csa climates are dry during the summer and Cfa climates are wet. Why?

What global pressure cells influence these regions during the winter?

G. The Alaskan pipeline is largely above ground. Part of the reason for this is for ease of maintenance. Can you think of another reason—more climatically related?

14.3 Learning Activities

A. Provide your students with a list of the thirty largest cities in the world. Have them determine the climatic region of each city. Have your students discuss the relationship (or lack of relationship) between climatic regions and world populations centers.

B. Have your students prepare a brief written summary in outline form of unique or outstanding climatic characteristics associated with the following climatic regions: humid continental, mild summer; subarctic; and tundra.

Name ______________________ Instructor ______________________

Date ______________________ Section Number ______________________

UNIT 15 - Biogeography

For use with Chapter 11

15.1 Natural Vegetation

Reference Pages in Text: Chapter 11, pp. 317-331

Materials Needed:

Student Supplied: colored pencils
Instructor Supplied: none

Purpose: This exercise will help you better understand the distribution of natural vegetation within the United States and the relationship of that distribution to climate.

A. On the following page outline maps are provided that highlight the natural vegetation and climatic regions of the continental United States. Determine a coding scheme (legend) and color-in the maps. Use Figures 9.4 and 11.21 as guides.

B. As Figure 11.21 and the maps you have just completed illustrate, there is considerable variation in the natural vegetation of the United States between 30 and 40 north latitude. Describe the broad changes in vegetation that occur within that latitudinal band as you move from the east coast of the United states to the west coast.

C. Discuss the climatic variation within that same latitudinal band. Does there appear to be a strong relationship between natural vegetation and climate?

Describe this relationship.

Vegetation

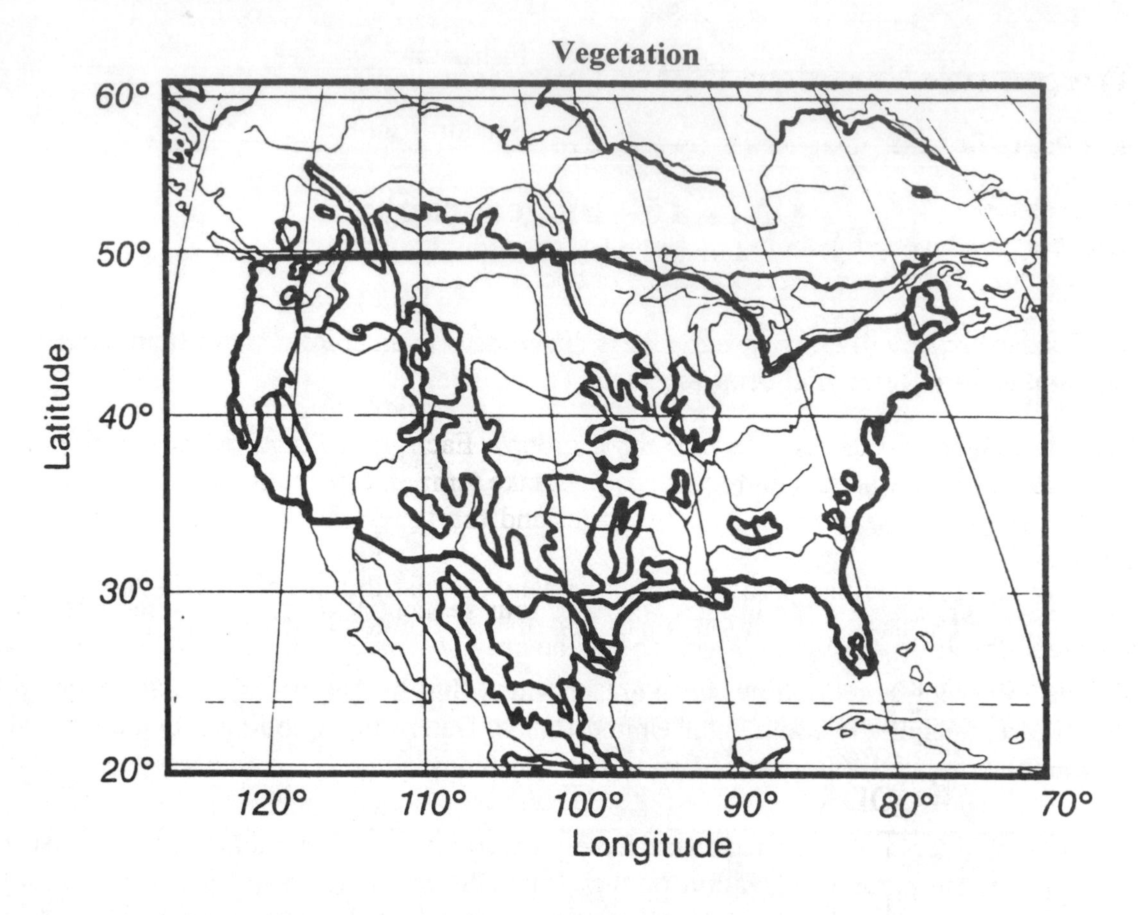

Climate

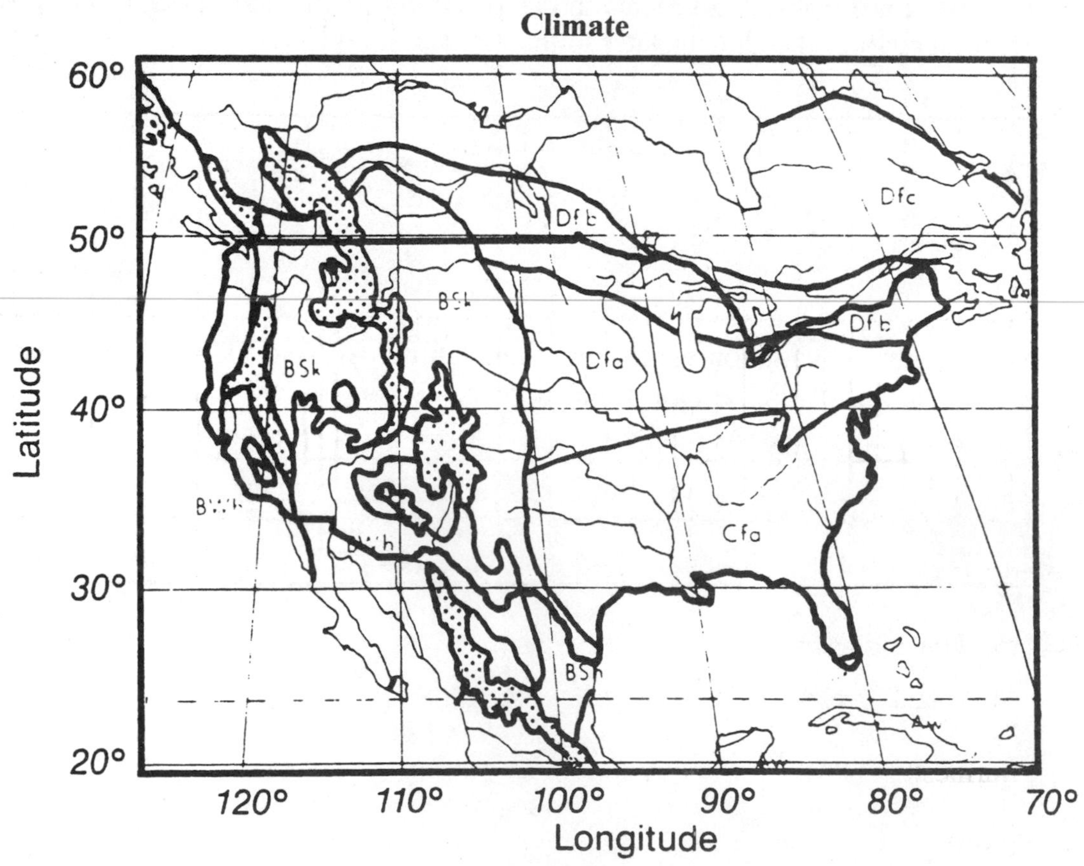

15.2 Temperature Variations Within Ecosystems

Reference Pages in Text: Chapter 11, pp. 309-314

Materials Needed:

Student Supplied: pencil, watch
Instructor Supplied: thermometers and meter sticks

Purpose: This exercise will allow you to assess differences in an abiotic factor (temperature) within and among three terrestrial ecosystems.

A. Your instructor will divide you into three groups. Each group must have a thermometer, a meter stick, and a watch. The first group will take temperature readings within a woods, the second group will take readings over grass, and the third group will take readings over a parking lot.

At predetermined times, the groups will take four sequential temperature readings: the first at ground level; the second at 30 cm above the ground surface; the third at 100 cm above the ground surface; and the last at 300 cm above the ground surface. Plot the results that the three groups obtained on the graph below.

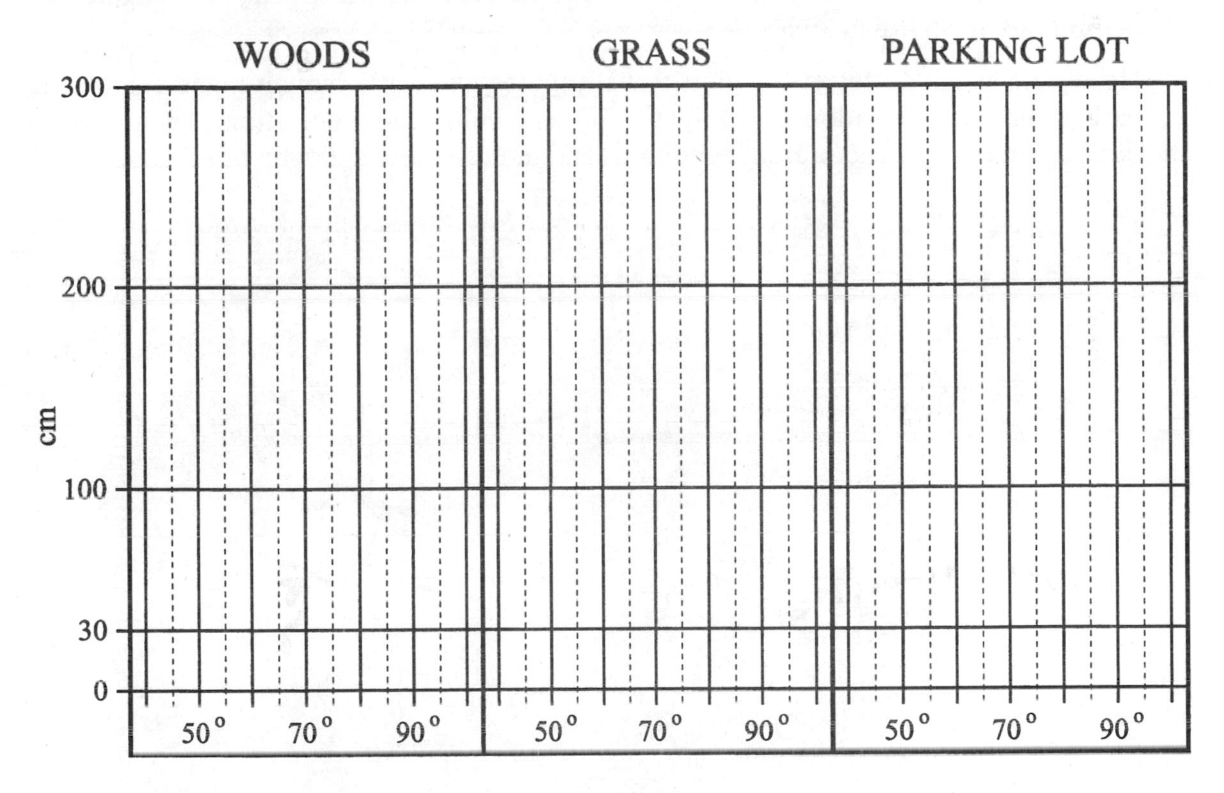

B. Study the data and answer the following questions:

1. At ground level which ecosystem is coolest?

   warmest?

   Why?

2. At which level above the ground are all three ecosystems most alike in temperature?

   Why?

3. How does the temperature in each ecosystem vary with elevation?

   Why?

15.3 Learning Activities

A. Although humans are the dominating organism in an urban area, they are obviously never the only organism. Have your students write an essay in which they investigate the biotic community in an urban area.

B. Have your students select a region that illustrates extensive human-made changes from an earlier natural vegetation cover to the present cultivated vegetation. Then, ask them to describe the earlier and present conditions and indicate how and why the change was made.

Name ______________________________ Instructor ____________________

Date ______________________________ Section Number ______________

UNIT 16 - Soil and Soil Characteristics

For use with Chapter 12

16.1 Determining Soil Texture

Reference Pages in Text: Chapter 12, pp. 340-341

Materials Needed:

Student Supplied: pencil
Instructor Supplied: none

Purpose: This exercise provides practice in identifying various soil textures by comparing the three basic particle sizes of the inorganic components of a soil; sand, silt, and clay. If the soil texture is known, the approximate percentages of sand, silt, and clay can be determined. These operations are accomplished by using the texture triangle illustrated in Figure 12.8 in the text.

A. When a soil sample is analyzed, the inorganic material is classed according to particle size. Sand particles are largest having diameters between 2.0mm and 0.05mm. Silt particles are next with diameters ranging between 0.05mm and 0.002mm. Clay particles, the smallest, have diameters less than 0.002mm. The percentage of sand, silt, and clay can be determined in a laboratory. Entering the texture triangle with these percentages will yield the soil texture. Care should be taken to insure the correct lines are followed in the triangle. This is clearly indicated by the orientation of the numbers along the borders of the triangle. The intersection of the lines that represent the percentages will occur in a texture area of the triangle, for example clay loam.

Using the texture triangle, determine the textures of the following soil samples.

	Sand	Silt	Clay	Texture
1.	35%	45%	20%	____________
2.	75%	15%	10%	____________
3.	10%	60%	30%	____________
4.	5%	45%	50%	____________

B. Conversely if the texture of a soil is known, the approximate percentages of the constituent particles can be determined. This necessitates using the texture triangle in reverse fashion from above.

What are the percentages of sand, silt, and clay of the following soil textures? (Note: the answers may vary, but the totals should add to 100%)

	Texture	Sand	Silt	Clay
1.	Sandy clay	_______	_______	_______
2.	Silty loam	_______	_______	_______

16.2 Determining Soil Porosity

Reference Pages in Text: Chapter 12, pp. 340-341

Materials Needed:
Student Supplied: pencil
Instructor Supplied: none

Purpose: This exercise introduces one of the procedures used to determine soil porosity.

A. Soil porosity is a physical characteristic that indicates the amount of space between soil particles and soil clumps. It is given in percentage of pore space. The porosity is important because it is a measure of how much water and air a particular soil can hold. Porosity can be determined by comparing the volume of a completely dry, soil core sample with the volume of water the core sample can hold when saturated. A common procedure for deriving the porosity is to weigh the dry core sample in grams, then weigh the water-saturated core sample in grams and subtract the dry weight from the saturated weight. This not only yields the weight of the water, but at the same time yields the volume in cubic centimeters because one cc of water weighs one gm. Next the volume of water is divided by the volume of the core sample. The steps are:

1. Weigh the dry core sample in grams.
2. Weigh the saturated core sample in grams.
3. Subtract #1 answer from #2 answer. This gives the cc of water.
4. Divide the cc of water by cc of the core sample and multiply by 100 to obtain the percentage of pore space which is the porosity.

B. Determine the following:

1. A 400cc core sample of oven-dry soil weighs 600 grams. When saturated with water this same sample weighs 760 grams. What is the percentage of pore space (porosity) of this soil?

2. A 400 cc core sample of oven-dry soil weighs 520 grams. When saturated with water this same sample weighs 720 grams. What is the porosity of this soil?

16.3 Soil Profiles

Reference Pages in Text: Chapter 12, pp 343-348 and pp. 352-361

Materials Needed:
Student Supplied: pencil
Instructor Supplied: none

Purpose: This exercise helps you gain familiarity with the characteristics of soil horizons in a soil profile and the processes operating in those horizons.

A. As the soil forms, distinctive horizontal layers known as soil horizons develop. These horizons develop over time as a result of the interaction of climate and organic activity with the parent material. The nature of these horizons determines the uses of a particular soil. One soil may be well suited for agriculture while another may be susceptible to erosion to the point where agricultural practices should be closely managed or avoided.

B. A generalized diagram of a soil profile is shown in Figure 12.12 in the text. Examine this figure and study the referenced pages to answer the following.

1. What horizons make up the zone of eluviation?

2. What are the two processes that occur in the zone of eluviation?

3. The various B horizons are in what zone?

4. Weathered parent material is the major constituent in what horizon?

5. Partly decomposed organic debris makes up which horizon?

B. Table 12.1 describes some common NRCS soil horizons. Study this table and the accompanying text to answer the following.

1. What materials accumulate in an argillic horizon?

2. What is an epipedon?

3. Generally, would a soil with an ochric epipedon or a mollic epipedon be better suited for agriculture?

4. What would be the name given to a seven inch thick horizon that contained at least two percent salt?

16.4 Temperature and Soil Formation

Reference Pages in Text: Chapter 12, pp. 345-354

Materials Needed:
Student Supplied: pencil
Instructor Supplied: none

Purpose: This exercise illustrates the fact that temperatures have an effect on the accumulation of humus which in turn is a very important component of a soil.

A. Refer to Figure 12.16 and the related text to answer these questions. Generally, the maximum accumulation of humus occurs within an optimum range of temperatures. Temperatures warmer or cooler than optimum result in a decrease of accumulation. Between what degrees of Fahrenheit temperatures does humus accumulate in the greatest amounts? (Circle the answer)

   Between 50 and 60   Between 60 and 70   Between 70 and 80   Between 80 and 90

B. What causes the humus accumulation to decrease at warmer than optimum temperatures?

C. What causes the humus accumulation to decrease at cooler than optimum temperatures?

16.5 Learning Activity

A. Have your students dig soil pits at a variety of locations within several miles of your school. Try to find undisturbed land rather than cultivated land and get permission to dig. These pits need only be a foot in diameter and a couple of feet deep. Have them note if there is any change in the character of the soil from the surface to the bottom of the pit. Some changes to check for are color, texture, moisture, and organic material. If possible, determine the percentages of sand, silt, and clay and classify the texture of the soil. With this data try to determine the main soil order using the descriptions of the soil orders found in the textbook. Then obtain a soil map of your area and compare the students' classifications with the soil map of your area.

Name ________________________________ Instructor ____________________

Date ________________________________ Section Number ______________

UNIT 21 - Tectonic Processes

For use with Chapter 14

21.1 Structural Features Related to Tectonic Processes

Reference Pages in Text: Chapter 14, pp. 402-409

Materials Needed:
Student Supplied: pencil
Instructor Supplied: physical wall map of North America

Purpose: This exercise will help you build an understanding of solid tectonic processes, including folding, warping, and faulting.

STRUCTURAL FEATURES

Structural features are those that result from deformed masses of rock; these include fractures, faults, and folds. Sedimentary rocks and some igneous rocks originate in approximately horizontal layers; when these are found tilted, folded, or faulted, they provide evidence of deformation of Earth's crust. Uplift and resulting erosion may strip away thousands of feet of the crust, revealing rocks that were once deeply buried.

In describing structural features, Earth scientists have found it convenient to use two special terms: dip and strike. If a rock layer is not horizontal, the amount of its slope is called dip, measured by the acute angle that the layer makes with the horizontal. The strike is the bearing of the outcrop of an inclined bed or structure on a level surface. A bed that has an east-west dip has a north-south strike. A bed that dips either to the north or to the south has an east-west strike.

FOLDS

Folds are a common feature of rock deformation produced by compressive stress. They range in size from microscopic to huge folds involving thousands of feet of rocks and distances of hundreds of miles. Folds may be tilted to one side and sometimes to one end (see the figure below). An Anticline is an upfold. A syncline is a downfold.

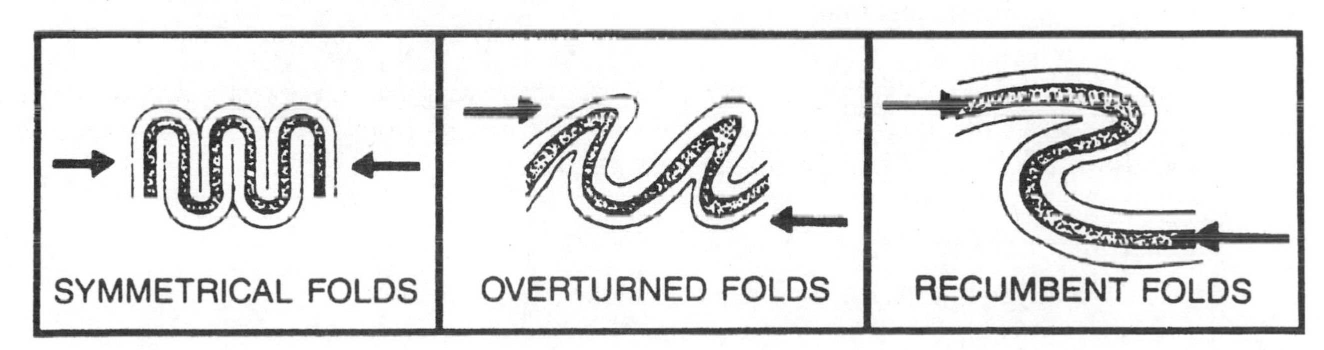

FAULTS

A fault occurs where masses of rock have moved or slipped past one another along a "crack" in Earth's surface. A fracture is a crack where no movement has taken place.

The basis for the classification of the faults is the nature of the relative movement of the rock masses on opposite sides of the fault. If displacement is in the direction of dip, it is called a Dip-Slip Fault. If displacement has been mainly horizontal, it is called a Strike-Slip Fault. The directions of movement involved in faulting are entirely relative; the absolute direction of movement usually cannot be determined. A thrust fault is a low-angle reverse fault. The diagrams below illustrate movement along a fault.

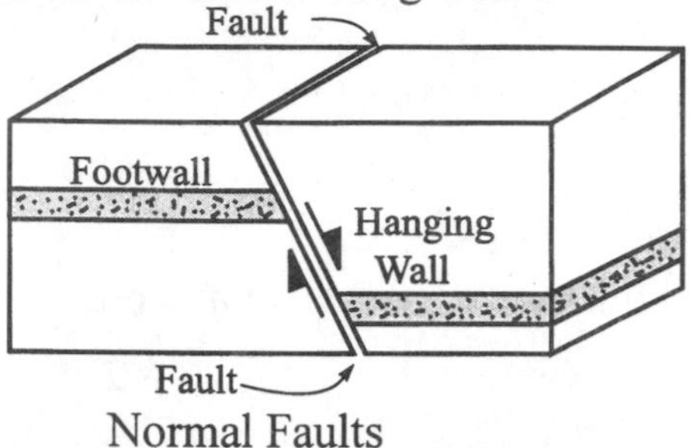

Normal Faults

A dip-slip fault in which the hanging wall has moved **down** in relationship to the footwall.

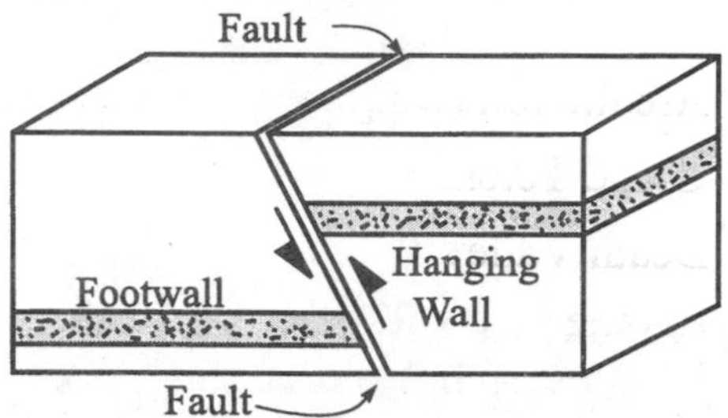

Reverse Faults

A dip-slip fault in which the hanging wall has moved **up** in relationship to the footwall.

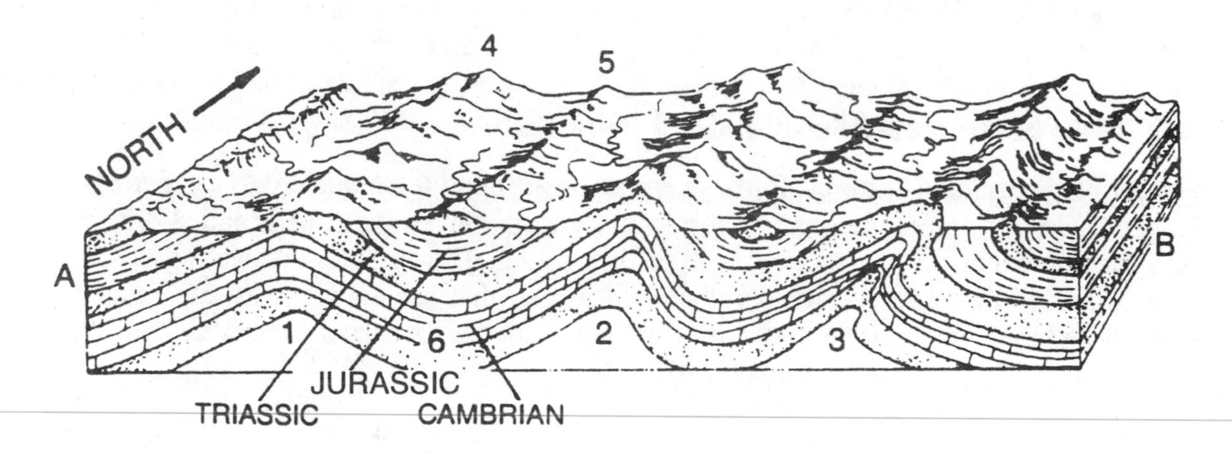

A. Identify the following features from the block diagram above and answer the questions that follow.

1. ___ symmetrical fold
2. ___ syncline
3. ___ recumbent fold
4. ___ anticlinal ridge
5. ___ synclinal ridge
6. ___ overturned fold
7. What type of topography is depicted in the above diagram?

8. Which mountain range in the United States would best show this type of topography?

9. The dip at "A" is
   and the strike is

10. The dip at "B" is
    and the strike is

11. Are the following landforms mainly due to folding, warping, or faulting?
    Grand Tetons
    Death Valley
    Florida
    Appalachians
    Sierra Nevada
    Colorado Plateau
    Basin and Range

21.2 Map Interpretation of Folded Mountains

Reference Pages in Text: Chapter 14, pp. 404 - 406, 414, see Figure 15.20., p. 428

Materials Needed:
Student supplied: pencil and straight edge
Instructor supplied: USGS Topographic Map of Harrisburg, Pennsylvania.
U.S. physical wall map.

Purpose: This exercise will help you build map interpretation skills in identifying folded landform features and the effects of stream erosion on folded structure.

A. Answer the following questions by using the topographic map of Harrisburg, Pennsylvania and the wall map of the United States.

1. Locate the Harrisburg quadrangle on a U.S. map. In what mountain range is it located?

2. The mountains are formed from resistant rock such as conglomerate and sandstone, while the valleys are formed from less resistant rock such as limestone and shale. Peters Mountain dips south, while Second Mountain dips north. Both mountain ridges are made of the same type of sedimentary rock. Third Mountain is much younger and made of resistant conglomerate. Based upon this information use the space at the top of the next page to diagram the folded structure.

3. What type of folded structure do Peters and Second Mountains form?

4. Third Mountain would form what type of surface feature?

5. What is the strike of the three mountain ridges?

6. In what direction is the Susquehanna River flowing?

   Into what water body does it empty?

7. What do we call a river such as the Susquehanna which cuts through resistant landforms?

21.3 LEARNING ACTIVITIES

A. .Have students answer questions similar to those in 21.2 in relation to topographic maps which show similar folding examples. Stereo-pair aerial photos also show faulted and folded structure and landforms well.

B. Have your students draw a series of diagrams illustrating both internal structure and external landforms produced by folding, warping, and faulting.

C. Have your students develop a collection of photographs which illustrate landforms produced by as wide a variety of different tectonic processes as possible. Students should attempt to include landforms produced in association with faulting, folding, and warping.

Name ______________________________ Instructor ____________________

Date ______________________________ Section Number ______________

UNIT 22 - Rocks

For use with Chapter 13

22.1 Rock Identification

Reference Pages in Text: Chapter 13, pp. 366 – 373, Appendix D

Materials needed:

Student Supplied: pencil

Instructor Supplied: samples of igneous, sedimentary, and metamorphic rocks; hand lens or magnifying glass; small bottles of HCL (for identifying limestone, dolomite, and calcite)

Purpose: This exercise will help you build an understanding of rock classification and the ability to identify various rock types.

A basic knowledge of rocks, the building blocks of the continents and ocean basins, will help you better understand the various landform processes and landform variety. Further, an understanding of rocks is important to the study of landforms and natural hazards. Some rocks are exceedingly strong and will withstand earthquakes. Others which are subject to faulting or mass movement can be extremely dangerous areas for human occupation.

A. Igneous Rocks

The definitions presented below pertain to igneous rocks. Study them in conjunction with Chart 1, which is presented at the end of this unit.

Definitions

Texture: size, shape, and arrangement of the mineral grains.

1. Coarse-grained texture: Most of the minerals in rock can be identified by the unaided eye.
2. Fine-grained texture: Individual rock forming minerals are for the most part too small to be recognized with the unaided eye.
3. Glass texture: Appearance of glass.

Fragmental: in large part of rock fragments. Pyroclastic: fragments (tephra) blown out by explosive volcanic eruptions and subsequently deposited on the ground where they weld together due to their latent heat. Includes ash, cinders, blocks, bombs, and pumice. (Rock name Ignimbrite.)

Porphyritic conspicuously large crystals (phenocrysts) occurring in a coarse or fine-grained matrix.

Vesicular: small cavities (vesicles) in a fine grained or glassy igneous rock, formed by the expansion of a bubble of gas during the solidification of the rock.

1. Scoriaceous: large number of vesicles. (Scoria)
2. Pumiceous: extreme number of vesicles.

Amygdaloidal: vesicles filled or partly filled with minerals at a later time.

Igneous rocks are classified based upon their texture and composition. Use Chart 1 to help you identify the igneous rock samples provided by your instructor.

NAME

Sample 1 ______________________

Sample 2 ______________________

Sample 3 ______________________

Sample 4 ______________________

Sample 5 ______________________

B. Sedimentary Rocks

The definitions presented below pertain to sedimentary rocks. Study them in conjunction with Charts 2 and 3 which are presented at the end of this unit.

Definitions

Clastic Sedimentary Rock: (Detrital) Rocks formed from accumulations of minerals and rocks; derived either from erosion of previously existing rock or from the weathered products of these rocks.

Biochemical Sedimentary Rock: A sedimentary rock made up of deposits resulting directly or indirectly from the life process of organisms.

Chemical Sedimentary Rock: A sedimentary rock composed chiefly of material deposited by chemical precipitation, either organic or inorganic. Compare with detrital sedimentary rock. Chemical sedimentary rocks may have either a clastic or nonclastic (usually crystalline) texture.

Oolite: Spheroidal grains of sand size, usually composed of calcium carbonate (calcite) and thought to have originated by inorganic precipitation. Some limestone is made largely of oolites.

Lithification: The process by which unconsolidated rock - forming materials are converted into a consolidated or coherent rock.

Fossil: Evidence of past life such as the bones of a dinosaur, shells of ancient clams, the footprint of a long-extinct animal, or the impression of a leaf in a rock.

Breccia: More than 50 percent of the inclusions are angular.

Conglomerate: More than 50 percent of the inclusions are rounded.

Some sedimentary rocks are classified by whether they are composed of broken bits and pieces of larger rocks or other debris (clastic or detrital). The common cementing agents in clastic sedimentary rocks are: quartz, calcite, iron-oxide (hematite and limonite) and clay minerals. Other sedimentary rocks have been formed from precipitation out of water (CHEMICAL).

Use Charts 2 and 3 to help you identify the sedimentary rock samples provided by your instructor.

NAME

Sample 1 ____________________

Sample 2 ____________________

Sample 3 ____________________

Sample 4 ____________________

Sample 5 ____________________

C. Metamorphic Rocks

The definitions presented below pertain to metamorphic rocks. Study them in conjunction with Chart 4, which is presented at the end of this unit.

Definitions

Texture: The position and arrangement of minerals or rocks.

Foliation: Layering in some rocks caused by parallel alignment of minerals. Produces rock cleavage.

Slaty Structure: Splits into smooth, flat plates.

Phyllitic Structure: Finely foliated; less perfectly developed than in slate; split with wavy fracture surface.

Schistose Structure: Close-packed, parallel orientation of minerals.

Gneissic Structure: Banded or layered, alternating light and dark colored minerals.

Metamorphic rocks have been heated and subjected to varying amounts of pressure and are by far the most complex of all rocks. Use Chart 4 to help you identify the metamorphic rock samples provided by your instructor.

NAME

Sample 1 ____________________

Sample 2 ____________________

Sample 3 ____________________

Sample 4 ____________________

Sample 5 ____________________

22.2 Learning Activities

A. Plan a local field trip visiting sites with different types of bedrock. Have the students correlate the rock type with the landforms, vegetation, soil, and land use at these same sites.

B. Obtain a geologic map of your state (these can be obtained from most state geological or mining offices). Have the students correlate the bedrock type with the landforms, soil, and land use of the state.

C. Have students collect various rock types from the local area and identify them.

Chart 1. Classification of Igneous Rocks

	Minerals	Orthoclase, Quartz. Muscovite, Biotite, Plagioclase, Hornblende	Plagioclase, Biotite, Hornblende, Augite	Plagioclase, Olivine, Augite
Intrusive		Light Colored	Intermediate Colored	Dark Colored
Intrusive	Coarse Grained	Granite	Diorite	Gabbro
Intrusive	Coarse Grained	Granodiorite	Granodiorite	
Extrusive	Fine Grained	Rhyolite	Andesite	Basalt
Extrusive	Vesicular Glassy	Pumice		
Extrusive	Glassy	Obsidian		
Extrusive	Vesicular	Scoria/Vesicular Basalt		
Extrusive	Pyroclastic: Coarse	Volcanic Breccia		
Extrusive	Pyroclastic: Fine	Tuff		

Chart 2. Clastic (Detrital) Sedimentary Rocks

Sediment Size	Composition	Features	Name	
Gravel	Any	Concrete look: 50% clasts, rounded.	Conglomerate	
Gravel	Any	Concrete look: 50% clasts, angular.	Breccia	
Sand	Any	Sand grains cemented together; any color.	Sandstone	
Silt	Any	Grains finer than sand but coarser than dust; any color.	Shale	Siltstone
Clay	Clay minerals	Smooth feeling with parallel fracture.		Mudstone
All sizes; angular	Any	Large angular fragments in copious amounts of clay.	Tillite	
Any	$CaCO_3$ shells	Broken bits and pieces of shells cemented together; effervescent.	Coquina	
Microscopic	SiO_2 skeletons of diatoms	Smooth, silky feel; looks like chalk.	Diatomite	
Microscopic	$CaCO_3$ skeletons	Chalky color, gritty feel, effervescent	Chalk	

Chart 3. Chemical Sedimentary Rocks

Name	Features	Composition
Limestone	cite hardness and effervescence; any color but usually light colored. May have fossils; FOSSILIFEROUS LIMESTONE.	Calcite ($CaCO_3$)
Chert	Hardness (7), conchoidal fracture. White to gray—CHERT; dark gray to black—FLINT; red—JASPER; green—JADITE. May have fossils: FOSSILIFEROUS CHERT.	Silica
Coal	Plant remains, black colored, light weight; will burn. Peat, lignite, bituminous, and anthracite are varieties.	Carbon
Gypsum	Soft (2), white or light colored.	Gypsum
Halite	"Rock" salt.	Salt (NaCl)

Chart 4. Classification of Metamorphic Rocks

Metamorphic Rocks							
Rocks commonly derived from	Structure	Texture	Visible Minerals	Color	Other Features	Rock Name	Meta grade
Shale	Foliated	Fine grained		variable	Flat, relatively dull fracture surfaces; breaks in smooth plates.	Slate	low
Shale			mica	variable	"Wavy" surface (breaks in plates).	Phyllite	
Shale, siltstones and basalts		Coarse grained	mica, garnet, quartz, hornblende	variable	Mica very conspicuous; garnet bead-like; finely foliated.	Schist	med.
Sedimentary rocks, igneous rocks			feldspar, quartz, mica, hornblende, augite	variable; white-gray	Generally banded, but foliation is visible; breaks generally in blocks.	Gneiss	high
Quartz, sandstone, chert	Non-foliated	Medium grained	quartz	variable	Interlocking grains with breakage across grains.	Quartzite	all
Limestone		Coarse grained	calcite	variable	Crystalline masses; will effervesce.	Marble	all

Name ______________________________ Instructor ____________________

Date ______________________________ Section Number ______________

UNIT 23 - Underground Water

For use with Chapter 16

23.1 Map Interpretation of Karst Topography

Reference Pages in Text: Chapter 17, pp. 451 - 461

Materials needed:

Student Supplied: pencils and straight edge

Instructor Supplied: U.S.G.S. Topographic Map of Interlachen, Florida and a United States wall map.

Purpose: This exercise will help you build map interpretation skills in identifying karst landforms.

A. Answer the following questions by using the topographic map of Interlachen, Florida.

1. As what type of major landform surface would this area be classified?

2. What is the contour interval of this map?

   Why do you think the cartographers chose this interval?

3. In what part of Florida is this area located?

4. This area is a region of karst topography. What is the derivation of the term "karst?"

5. On what type of bedrock is the region situated?

6. From the map, how do you know this is a karst region?

7. What is the approximate elevation of the water table? (Note: You can determine this from the elevation of the lakes.)

8. Underground, the water flows through an aquifer. Define "aquifer" and list characteristics an aquifer must have.

9. Based on information from the map, in what direction would groundwater flow?

10. As Florida is a rapidly-urbanizing state, what problems and hazards do you foresee for this karst area?

    a.

    b.

    c.

    d.

23.2 Learning Activities

A. Have the students answer questions similar to those in 23.1 in relation to topographic maps of other karst regions. Suggested USGS quadrangles:

   1. Mammoth Cave, Kentucky
   2. Sulphur Springs, Florida
   3. Bottomless Lakes, New Mexico
   4. Ashby, Nebraska

B. Plan a field trip if you are located in a region which has cave and cavern formations. Most public (state and national park) and privately-owned caves have brochures and maps available to plan a lecture prior to your visit.

C. Have students write an essay on groundwater pollution. In the essay have them address the subject of septic tanks, explaining how they function and the problem of improper maintenance.

Name ______________________________ Instructor ____________________

Date ______________________________ Section Number ______________

UNIT 24 - Surface Water

For use with Chapter 17

24.1 Surface Water Geography

References Pages in Text: Chapter 17, pp. 466, 483 - 486

Materials Needed:
Student Supplied: blue and red pencils
Instructor Supplied: physical world wall map or atlas

Purpose: This exercise will help you build your geographic knowledge of the world's major surface water features.

A. Major Rivers

1. On the appropriate world region maps (pages 105 and 106) that follow, draw the following rivers in blue pencil. Label them by the letters indicated.

South America
a. Amazon
b. Paraná
c. Orinóco

North America
i. Mississippi-Missouri
j. St. Lawrence
k. Colorado
l. Columbia

Australia
d. Murray-Darling

Africa
e. Congo
f. Nile
g. Niger
h. Zambezi

Eurasia
m. Rhine
n. Rhone
o. Danube
p. Volga
q. Tigris-Euphrates
r. Indus
s. Ganges
t. Brahmaputra
u. Mekong
v. Chang Jiang (Yangtze)
w. Huang He (Yellow)
x. Ob
y. Yenisey
z. Lena

2. Which river has the largest drainage basin and greatest annual discharge?

3. Which river is considered the longest river?

4. Which river system has North America's largest drainage basin and greatest annual flow?

B. Major Lakes

1. On the world regional maps that follow, draw the following lakes in red pencil. Label them by the letters indicated.
   a. Caspian Sea
   b. Lake Superior
   c. Lake Victoria
   d. Aral Sea
   e. Lake Huron
   f. Lake Michigan
   g. Lake Tanganyika
   h. Lake Baikal

2. What is the world's largest lake?

3. What is the world's deepest lake?

4. What is the world's largest interconnected lake system?

5. Why are some lakes called "seas?"

6. Name two large salty lakes.

7. What is the origin of the basin occupied by the following water bodies?
   a. Crater Lake, Oregon

   b. Lake Baikal, USSR

   c. Minnesota's 10,000 lakes

   d. The Great Lakes

   e. Lake Nasser, Egypt

   f. Lake Mead, Arizona-Nevada

   g. Lake Tahoe, California

   h. The Dead Sea, Israel-Jordan

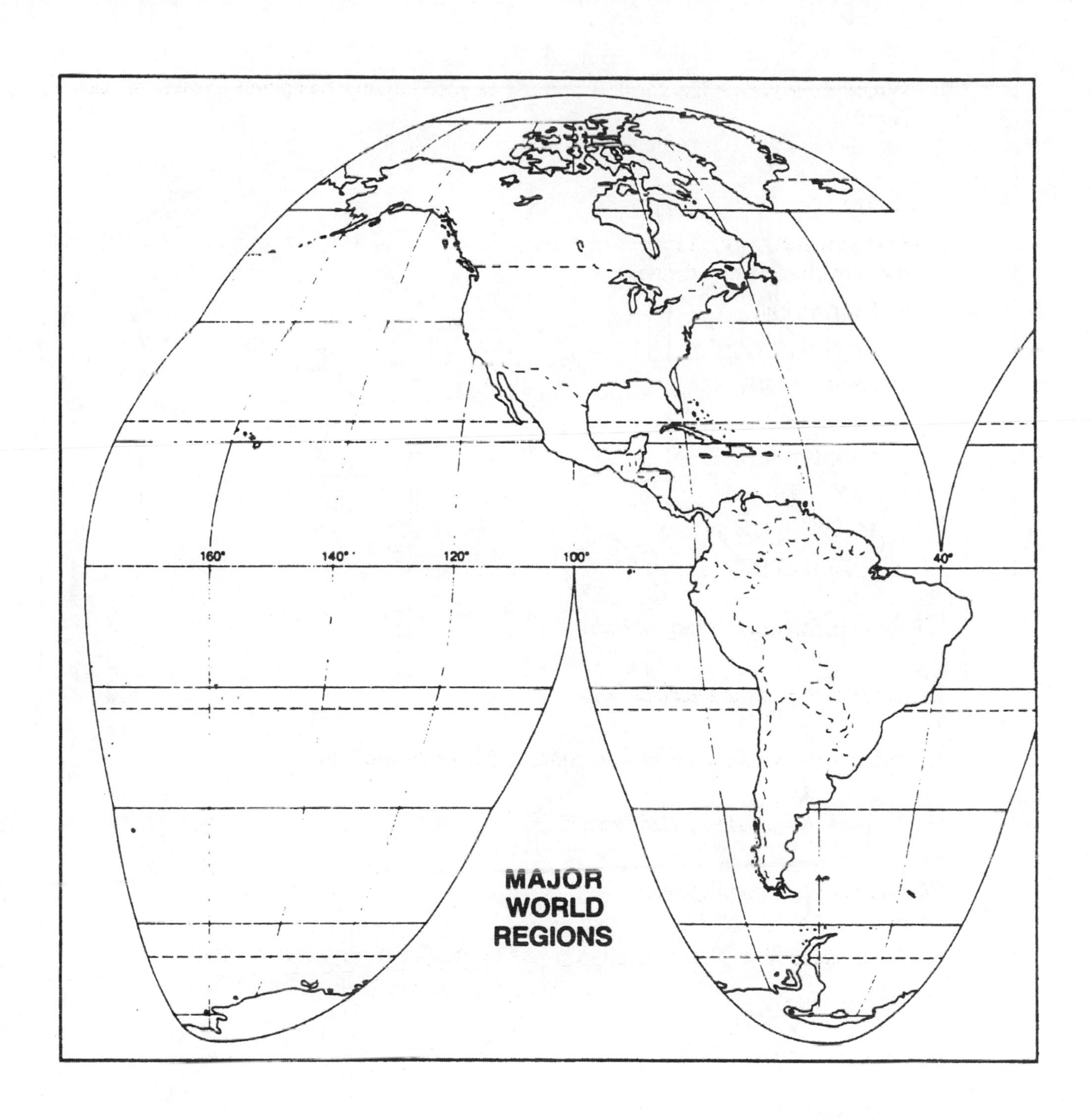
160°
140°
120°
100°
40°
MAJOR
WORLD
REGIONS

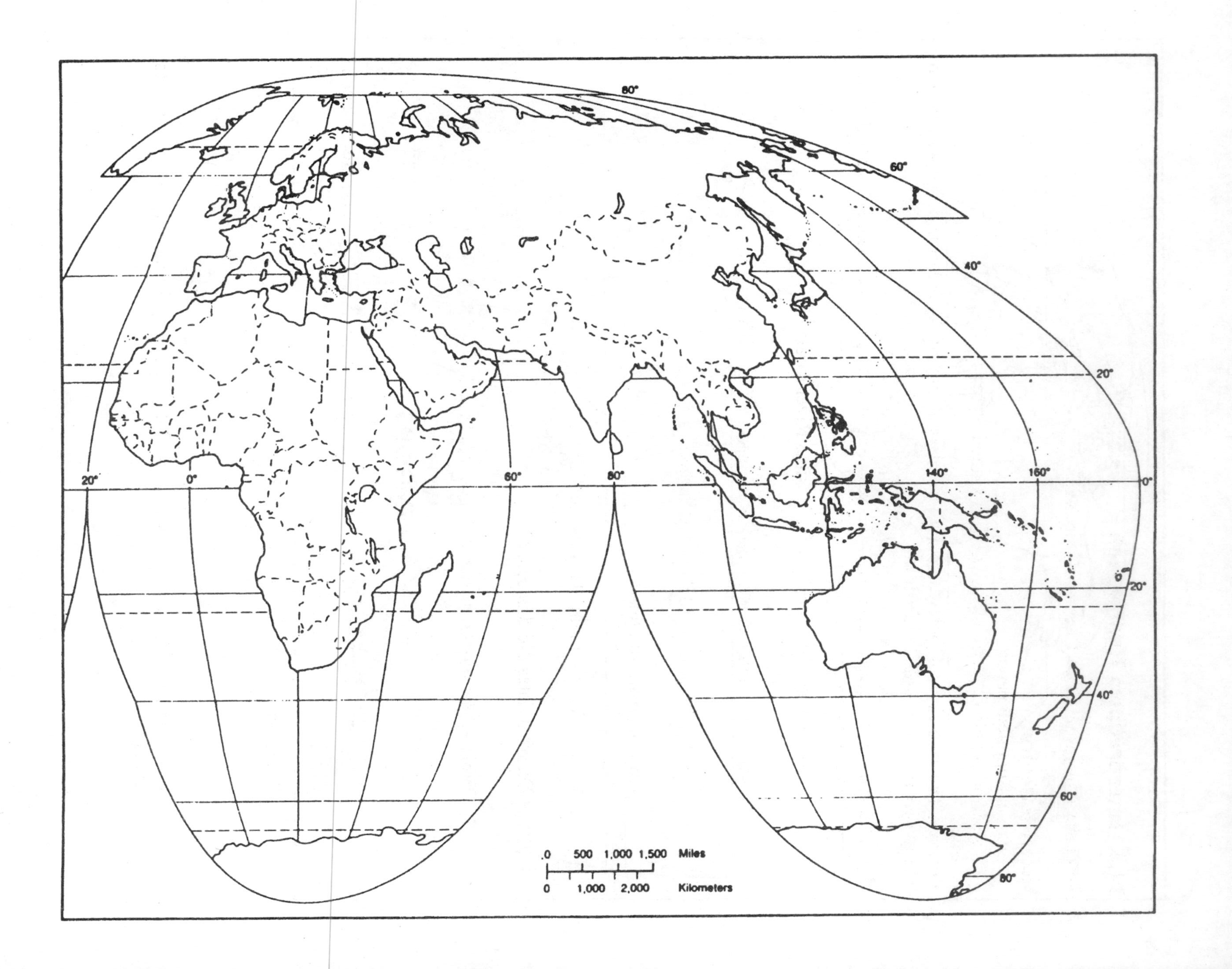
80°
60°
40°
20°
20°
0°
60°
80°
140°
160°
0°
20°
40°
60°
80°
0 500 1,000 1,500 Miles
0 1,000 2,000 Kilometers

24.2 Interpreting a Hydrograph

References pages in Text: Chapter 17, pp. 462 - 465

Materials Needed:
Student Supplied: pencil
Instructor Supplied: physical wall map of United States

Purpose: This exercise will help you build an understanding of the use of hydrographic data to interpret characteristics of stream flow.

A. Using the hydrograph (page 108) for the Yakima River at Union Gap, Washington, answer the following questions:

1. In what major river basin is the Yakima River located?

2. What mountain range is partially drained by the Yakima River?

3. What is the maximum daily discharge?

   On what dates did this occur?

4. What is the minimum daily discharge?

   On what dates did this occur?

5. Which three months had the greatest discharge?

6. Which three months had the minimum discharge?

7. Do the maximum and minimum discharges coincide with the general precipitation pattern for this region?

   If not, why?

8. How might you explain the series of peak flows during the months of November to March?

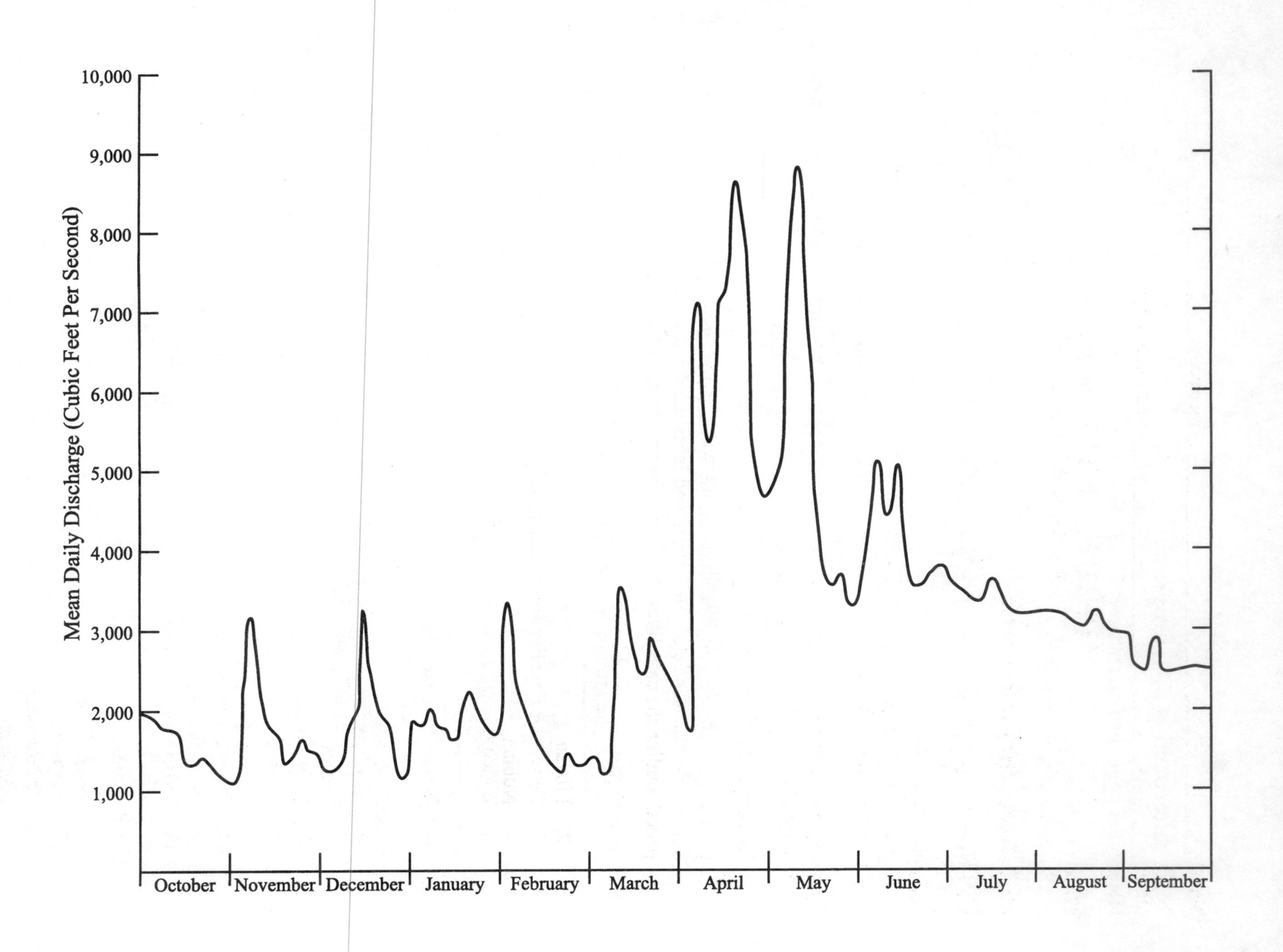
Mean Daily Discharge (Cubic Feet Per Second)
10,000
9,000
8,000
7,000
6,000
5,000
4,000
3,000
2,000
1,000
October
November
December
January
February
March
April
May
June
July
August
September

Name ______________________________ Instructor ____________________

Date ______________________________ Section Number ______________

UNIT 25 - Arid Landforms

For use with Chapter 18

25.1 Geography of Deserts

Reference Pages in Text: Chapter 18, general

Materials Needed:

Student Supplied: red and blue pencils

Instructor Supplied: world wall map and world climate map and/or atlas

Purpose: This exercise will help you build your geographic knowledge of the distribution of the world's deserts.

A. Deserts of the World

Locate the following major deserts on the world maps (pages 111 and 112). In red pencil, outline the boundaries of the deserts and label them with the appropriate number. In blue pencil, outline the coastlines of the cool current deserts (*).

Africa: Sahara

1. Western Sahara*
2. Libyan
3. Eastern or Egyptian (also called Arabian)
4. Nubian
5. Somali
6. Kalahari
7. Namib*

Eurasia: Arabian

8. Ar Rub al Khali (The Empty Quarter)
9. Syrian
10. Negev-Sinai
11. Iranian
12. Turkestan
13. Thar (Great Indian)
14. Gobi (Shamo)
15. Takla Makan (Tarim)

North America

16. Great Basin
17. Mojave
18. Chihuahuan
19. Sonoran
20. Baja California*

South American

21. Patagonia*
22. Atacama-Peruvian*

Australia

23. Great Victoria
24. Gibson
25. Great Sandy
26. Simpson

B. Answer the following questions relating to the world's deserts:

1. Which is the world's largest desert:

2. Which is the world's smallest desert:

3. Most of the world's deserts are centered at about:

4. Name two deserts which are mainly plains:

5. Name two deserts which are mountainous:

6. Name two deserts which are located in the leeward "rain shadow" of a high mounta range:

7. Which continent has the greatest percentage of its area in desert?

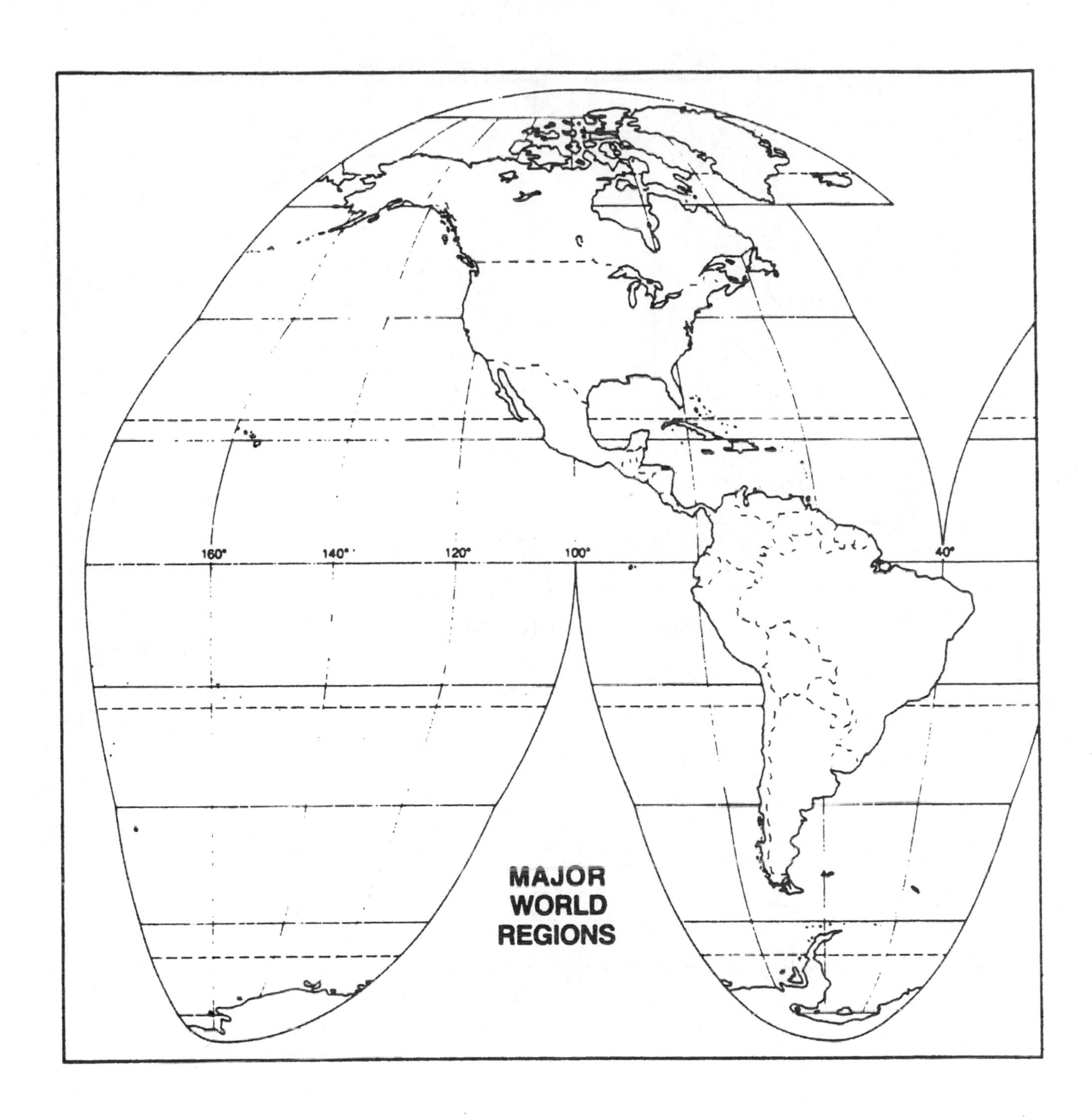
160°
140°
120°
100°
40°
MAJOR
WORLD
REGIONS

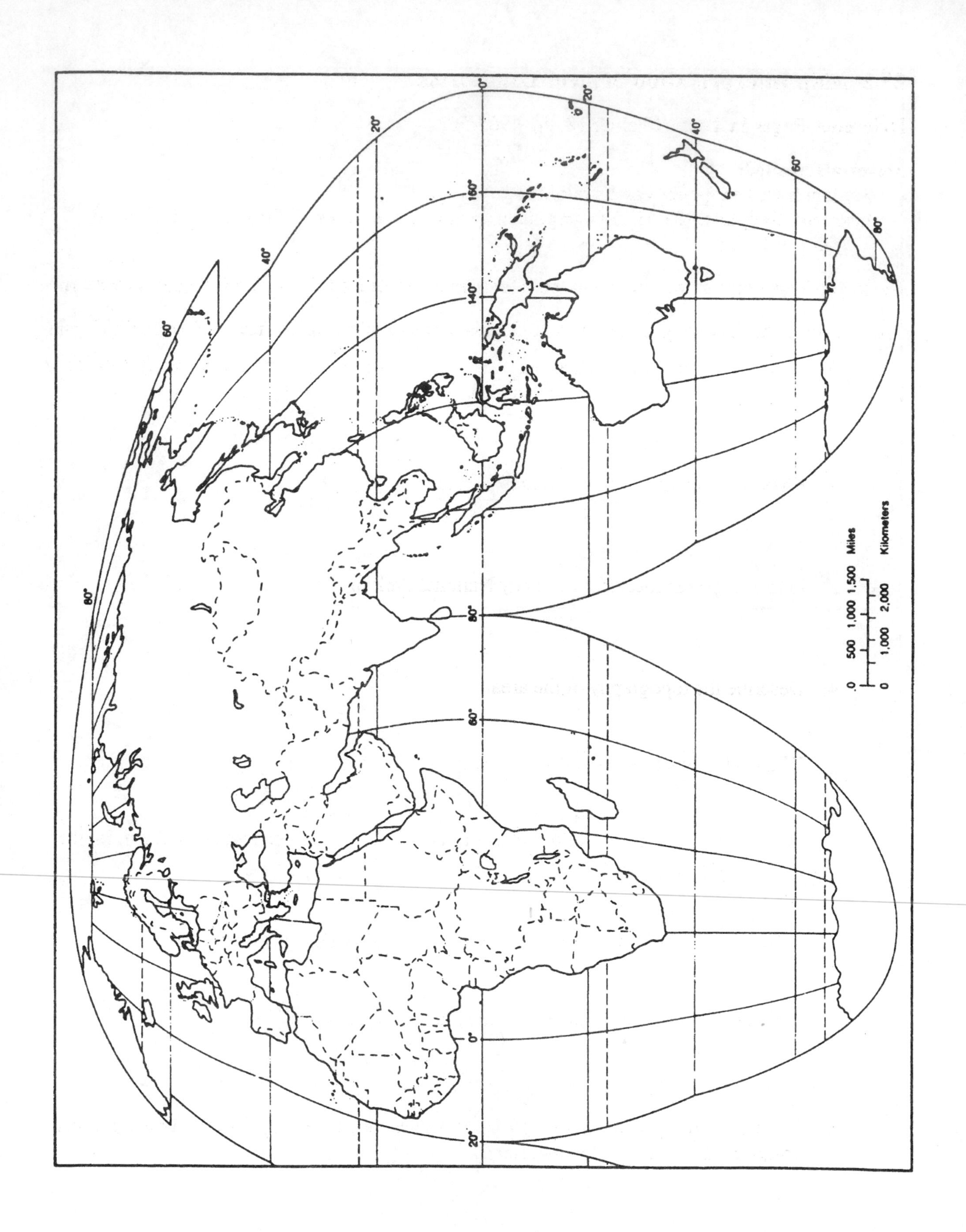

0 500 1,000 1,500 Miles
0 1,000 2,000 Kilometers

25.2 Map Interpretation of Arid Landforms

Reference Pages in Text: Chapter 18, pp. 490 - 502

Materials needed:

Student Supplied: pencils and straight edge
Instructor Supplied: USGS Topographic Map of Furnace Creek, California, and a United States wall map

Purpose: This exercise will help you build map interpretation skills in identifying arid landforms.

A. Answer the following questions by using the topographic map of Furnace Creek, California.

1. Locate the Furnace Creek area on a U.S. map. In what part of California is it located?

2. Based on location, why is this area so arid?

3. Is this map area located within any National Park or Monument? Which one?

4. Describe the topography of the area.

5. What is the lowest elevation on the map?

6. What specific type of arid landform is the blue striped feature located in the depression?

   Why are the edges of this feature shown with blue dashed lines?

   What type of surface materials is this feature composed of?

7. What specific type of arid landform is indicated on the map by the curved parallel contours at the base of the mountains?

What type of surface materials is this feature composed of?

What do the blue — ·· — ·· lines crossing the contours indicate?

Note that the large curved landforms at the base of the mountains coalesce into other similar features. What is the specific landform name for such broad features?

8. What evidence from the map indicates that this is an interior drainage basin?

B. Construct an east-west profile from the benchmark at Devil's Speedway to the 2,389 foot benchmark on the mountain straight to the west. Label the following features: mountain front, pediment, alluvial fan, and basin.

25.3 Learning Activities

A. Have your students write an essay in which they offer detailed reasons for the "Dust Bowl" in the American Midwest during the 1930s.

B. Have your students discuss methods that might be used to abate the desert encroachment currently occurring in the Sahel.

C. Have your students collect pictures from magazines depicting as many desert landforms as possible.

Name ______________________________ Instructor ________________

Date ______________________________ Section Number ____________

UNIT 28 – The Global Ocean

For use with Chapter 20

28.1 Marine Geography

Reference Pages in Text: Chapter 20

Materials Needed:
Student Supplied: pencil, calculator, straight edge
Instructor Supplied: world wall map and world atlas

Purpose: This exercise will help you build your geographic knowledge of the world ocean.

A. On the world outline maps that follow (page 125 and 126), label the world ocean features listed below.

PACIFIC OCEAN

1. Bering Sea
2. Sea of Okhotsk
3. Sea of Japan
4. Yellow Sea
5. East China Sea
6. South China Sea
7. Ross Sea (in Antarctic)
8. Java Sea
9. Arafura Sea
10. Tasman Sea
11. Philippine Sea
12. Gulf of Alaska
13. Gulf of California
14. Coral Sea

ATLANTIC OCEAN

15. Mediterranean Sea
16. Black Sea
17. North Sea
18. Norwegian Sea
19. Baltic Sea
20. Gulf of Guinea
21. Weddell Sea (in Antarctic)
22. Baffin Bay
23. Hudson Bay
24. Labrador Sea
25. Greenland Sea
26. Gulf of Mexico
27. Caribbean Sea
28. Sargasso "Sea"

INDIAN OCEAN

29. Arabian Sea
30. Bay of Bengal
31. Red Sea
32. Persian Gulf

ARCTIC "OCEAN"

33. Barents Sea
34. Beaufort Sea
35. Laptev Sea
36. Chukchi Sea

B. The surface areas of Earth's oceans and major seas are presented below.

Pacific Ocean	63,800,000 mi^2
Atlantic Ocean	31,800,000 mi^2
Indian Ocean	28,400,000 mi^2
Arctic Ocean	5,400,000 mi^2
Caribbean Sea	1,063,000 mi^2
Mediterranean Sea	967,000 mi^2
Bering Sea	876,000 mi^2
Gulf of Mexico	596,000 mi^2

On the graph below, the four-inch-long bar represents the surface area of the Pacific Ocean. Complete the graph by drawing correctly-scaled bars for the other oceans and seas.

Pacific Ocean	4"
Atlantic Ocean	
Indian Ocean	
Arctic Ocean	
Caribbean Sea	
Mediterranean Sea	
Bering Sea	
Gulf of Mexico	

28.2 Tides

Reference Pages in Text: Chapter 20, pp. 568 - 574

Materials Needed:
Student Supplied: pencil
Instructor Supplied: none

Purpose: This exercise demonstrates several facets of tides.

A. The table on page 129 lists water heights, at two-hour intervals over a twenty-four hour period. Plot this data on the graph below the table.

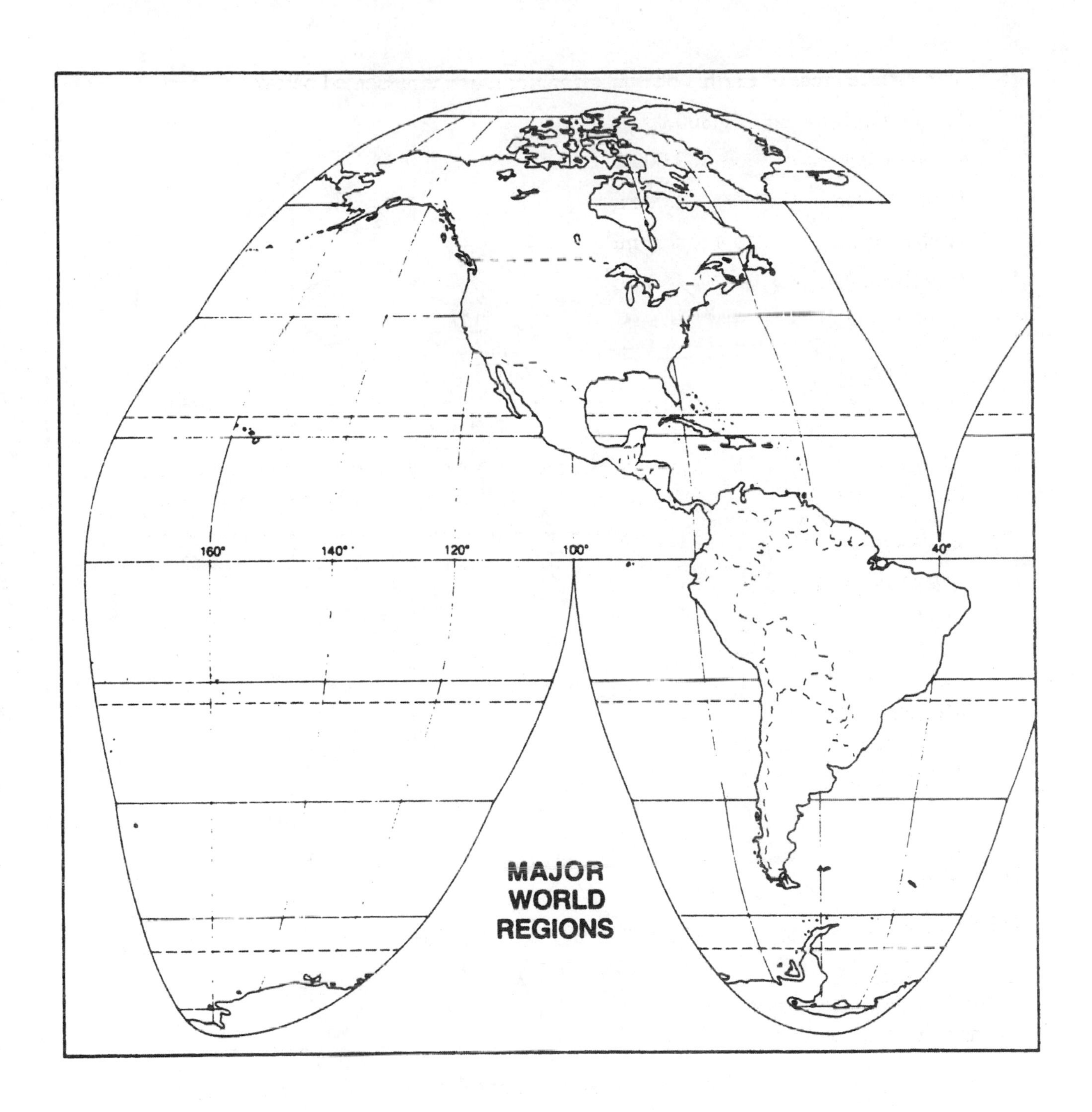
160°
140°
120°
100°
40°
MAJOR
WORLD
REGIONS

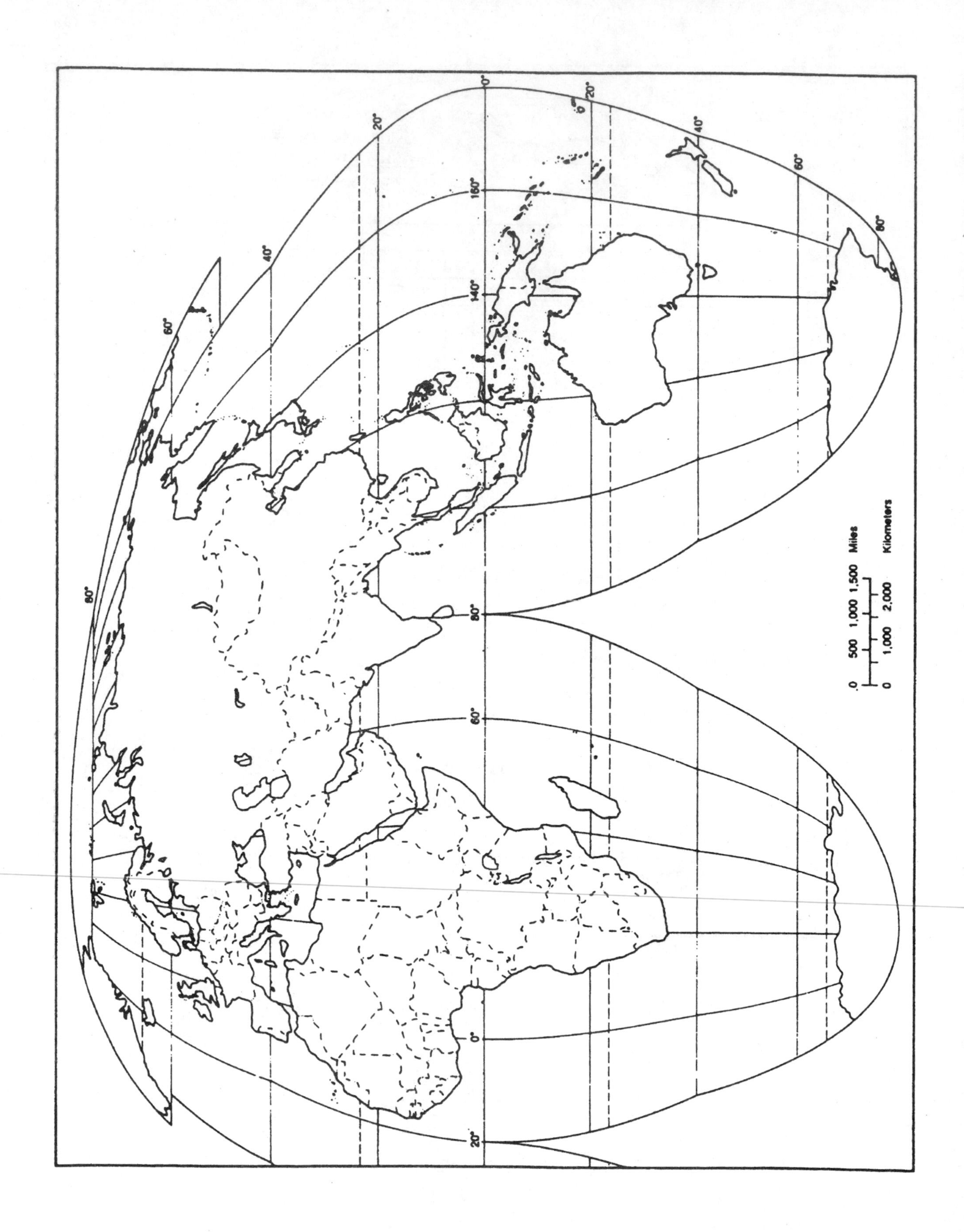
0 500 1,000 1,500 Miles
0 1,000 2,000 Kilometers
80°
60°
40°
20°
0°
20°
40°
60°
80°
160°
140°
80°
60°
0°
20°

Water heights

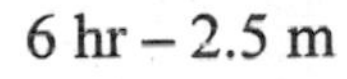

0 hr – 1.5 m
2 hr – 2.5 m
4 hr – 3.0 m
6 hr – 2.5 m
8 hr – 2.0 m
10 hr – 1.5 m
12 hr – 1.0 m
14 hr – 1.5 m
16 hr – 2.0 m
18 hr – 1.5 m
20 hr – 1.0 m
22 hr – 0.5 m
24 hr – 1.0 m

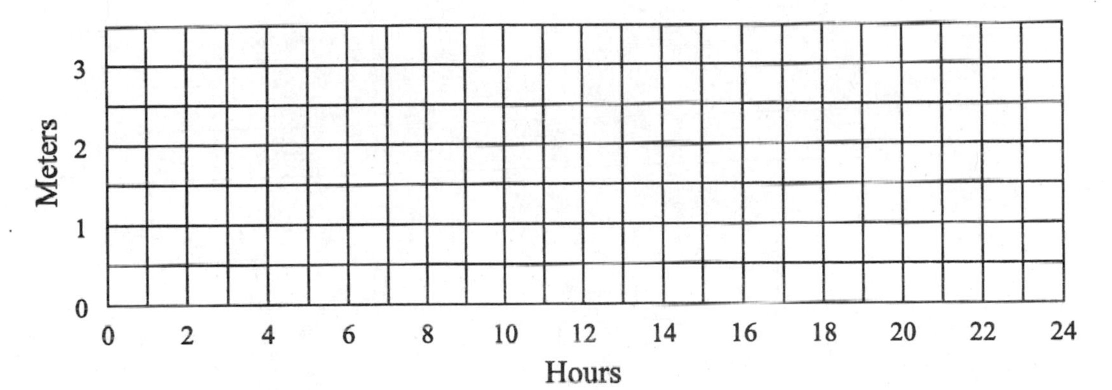

B. Answer the following questions:

1. What are the heights of the two high tides and two low tides?
   a. high tides _______ _______

   b. low tides _______ _______

2. What are the ranges between successive high and low tides:
   a. _______

   b. _______

What type of tidal curve do these data represent?

Why?